Algebra through practice

**Book 5: Groups**

# Algebra through practice

*A collection of problems in algebra with solutions*

# Book 5
# Groups

T. S. BLYTH o E. F. ROBERTSON

*University of St Andrews*

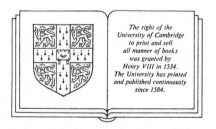

The right of the
University of Cambridge
to print and sell
all manner of books
was granted by
Henry VIII in 1534.
The University has printed
and published continuously
since 1584.

CAMBRIDGE UNIVERSITY PRESS

*Cambridge*

*London  New York  New Rochelle*

*Melbourne  Sydney*

CAMBRIDGE UNIVERSITY PRESS
Cambridge, New York, Melbourne, Madrid, Cape Town, Singapore, São Paulo

Cambridge University Press
The Edinburgh Building, Cambridge CB2 8RU, UK

Published in the United States of America by Cambridge University Press, New York

www.cambridge.org
Information on this title: www.cambridge.org/9780521272902

First published 1985
Re-issued in this digitally printed version 2008

A catalogue record for this publication is available from the British Library

Library of Congress Catalogue Card Number: 83–24013

ISBN 978-0-521-27290-2 paperback

# Contents

# Preface

The aim of this series of problem-solvers is to provide a selection of worked examples in algebra designed to supplement undergraduate algebra courses. We have attempted, mainly with the average student in mind, to produce a varied selection of exercises while incorporating a few of a more challenging nature. Although complete solutions are included, it is intended that these should be consulted by readers only after they have attempted the questions. In this way, it is hoped that the student will gain confidence in his or her approach to the art of problem-solving which, after all, is what mathematics is all about.

The problems, although arranged in chapters, have not been 'graded' within each chapter so that, if readers cannot do problem $n$ this should not discourage them from attempting problem $n+1$. A great many of the ideas involved in these problems have been used in examination papers of one sort or another. Some test papers (without solutions) are included at the end of each book; these contain questions based on the topics covered.

TSB, EFR
St Andrews

# Background reference material

Courses on abstract algebra can be very different in style and content. Likewise, textbooks recommended for these courses can vary enormously, not only in notation and exposition but also in their level of sophistication. Here is a list of some major texts that are widely used and to which the reader may refer for background material. The subject matter of these texts covers all six of the present volumes, and in some cases a great deal more. For the convenience of the reader there is given overleaf an indication of which parts of which of these texts are most relevant to the appropriate sections of this volume.

[1] I. T. Adamson, *Introduction to Field Theory*, Cambridge University Press, 1982.
[2] F. Ayres, Jr, *Modern Algebra*, Schaum's Outline Series, McGraw-Hill, 1965.
[3] D. Burton, *A first course in rings and ideals*, Addison-Wesley, 1970.
[4] P. M. Cohn, *Algebra* Vol. I, Wiley, 1982.
[5] D. T. Finkbeiner II, *Introduction to Matrices and Linear Transformations*, Freeman, 1978.
[6] R. Godement, *Algebra*, Kershaw, 1983.
[7] J. A. Green, *Sets and Groups*, Routledge and Kegan Paul, 1965.
[8] I. N. Herstein, *Topics in Algebra*, Wiley, 1977.
[9] K. Hoffman and R. Kunze, *Linear Algebra*, Prentice Hall, 1971.
[10] S. Lang, *Introduction to Linear Algebra*, Addison-Wesley, 1970.
[11] S. Lipschutz, *Linear Algebra*, Schaum's Outline Series, McGraw-Hill, 1974.

[12] I. D. Macdonald, *The Theory of Groups*, Oxford University Press, 1968.
[13] S. MacLane and G. Birkhoff, *Algebra*, Macmillan, 1968.
[14] N. H. McCoy, *Introduction to Modern Algebra*, Allyn and Bacon, 1975.
[15] J. J. Rotman, *The Theory of Groups: An Introduction*, Allyn and Bacon, 1973.
[16] I. Stewart, *Galois Theory*, Chapman and Hall, 1975.
[17] I. Stewart and D. Tall, *The Foundations of Mathematics*, Oxford University Press, 1977.

## References useful for Book 5

1: Subgroups   [**4**, Sections 9.1, 9.6], [**6**, Chapter 7],
[**8**, Sections 2.1, 2.11], [**12**, Chapters 1–6],
[**13**, Sections 13.1, 13.4], [**15** Chapters 1–4].
2: Automorphisms and Sylow theory   [**4**, Sections 9.4, 9.8],
[**8**, Section 2.12], [**12**, Chapter 7], [**13**, Section 13.5],
[**15**, Chapter 5].
3: Series   [**4**, Sections 9.2, 9.5], [**12**, Chapters 9,10],
[**13**, Sections 13.6–13.8], [**15**, Chapter 6].
4: Presentations   [**4**, Section 9.9], [**12**, Chapter 8],
[**15**, Chapter 11].

In [**8**] morphisms are written on the left but permutations are written as mappings on the right. In [**4**] and [**12**] all mappings (including permutations) are written as mappings on the right. In American texts 'solvable' is used where we have used 'soluble'.

# 1: Subgroups

The isomorphism and correspondence theorems for groups should be familiar to the reader. The first isomorphism theorem (that if $f : G \to H$ is a group morphism then $G/\operatorname{Ker} f \simeq \operatorname{Im} f$) is a fundamental result from which follow further isomorphisms : if $A \leq G$ (i.e. $A$ is a subgroup of $G$), if $N \triangleleft G$ (i.e. $N$ is a normal subgroup of $G$), and if $K \triangleleft G$ with $K \leq N$, then

$$A/(A \cap N) \simeq NA/N \quad \text{and} \quad G/N \simeq (G/K)/(N/K).$$

The correspondence theorem relates the subgroups of $G/N$ to the subgroups of $G$ that contain $N$.

Elements $a, b$ of $G$ are said to be conjugate if $a = g^{-1}bg$ for some $g \in G$. Conjugacy is an equivalence relation on $G$ and the corresponding classes are called conjugacy classes. The subset of $G$ consisting of those elements that belong to singleton conjugacy classes forms a normal subgroup $Z(G)$ called the centre of $G$. For $H \leq G$ the subset

$$\mathcal{N}_G(H) = \{g \in G \mid (\forall h \in H) \, g^{-1}hg \in H\}$$

is called the normaliser of $H$ in $G$. It is the largest subgroup of $G$ in which $H$ is normal. The derived group of $G$ is the subgroup $G'$ generated by all the commutators $[a, b] = a^{-1}b^{-1}ab$ in $G$, and is the smallest normal subgroup of $G$ with abelian quotient group.

Examples are most commonly constructed with groups of matrices (subgroups of the group $\operatorname{GL}(n, F)$ of invertible $n \times n$ matrices with entries in a field $F$), groups of permutations (subgroups of the symmetric groups $S_n$), groups given by generators and relations, and direct (cartesian) products of given groups.

An example of a presentation is

$$G = \langle\, a, b \mid a^2 = b^3 = 1,\ a^{-1}ba = b^{-1} \,\rangle.$$

Since $|\langle b \rangle| = 3$ and $\langle b \rangle \lhd G$ with $G/\langle b \rangle \simeq C_2$ (the cyclic group of order 2), we see that $|G| = 6$. The generators $a$ and $b$ can be taken to correspond to the permutations $(1\,2)$ and $(1\,2\,3)$ which generate $S_3$, or to the matrices

$$\begin{bmatrix} 0 & 1 \\ 1 & 0 \end{bmatrix}, \qquad \begin{bmatrix} 1 & 1 \\ 1 & 0 \end{bmatrix}$$

which generate $\mathrm{SL}(2, \mathbb{Z}_2)$, the group of $2 \times 2$ matrices of determinant 1 with entries in the field $\mathbb{Z}_2$. Thus we have that $G \simeq S_3 \simeq \mathrm{SL}(2, \mathbb{Z}_2)$.

**1.1**    Let $G$ be a group, let $H$ be a subgroup of $G$ and let $K$ be a subgroup of $H$. Prove that
$$|G : K| = |G : H|\,|H : K|.$$

Deduce that the intersection of a finite number of subgroups of finite index is a subgroup of finite index. Is the intersection of an infinite number of subgroups of finite index necessarily also of finite index?

**1.2**    Let $G$ be a group and let $H$ be a subgroup of $G$. Prove that the only left coset of $H$ in $G$ that is a subgroup of $G$ is $H$ itself. Prove that the assignment
$$\varphi : xH \mapsto Hx^{-1}$$

describes a mapping from the set of left cosets of $H$ in $G$ to the set of right cosets of $H$. Show also that $\varphi$ is a bijection. Does the prescription

$$\psi : xH \mapsto Hx$$

describe a mapping from the set of left cosets of $H$ to the set of right cosets of $H$? If so, is $\psi$ a bijection?

**1.3**    Find a group $G$ with subgroups $H$ and $K$ such that $HK$ is not a subgroup.

**1.4**    Consider the subgroup $H = \langle (1\,2) \rangle$ of $S_3$. Show how the left cosets of $H$ partition $S_3$. Show also how the right cosets of $H$ partition $S_3$. Deduce that $H$ is not a normal subgroup of $S_3$.

**1.5**    Let $G$ be a group and let $H$ be a subgroup of $G$. If $g \in G$ is such that $|\langle g \rangle| = n$ and $g^m \in H$ where $m$ and $n$ are coprime, show that $g \in H$.

**1.6**  Let $G$ be a group. Prove that

  (i) If $H$ is a subgroup of $G$ then $HH = H$.

  (ii) If $X$ is a finite subset of $G$ with $XX = X$ then $X$ is a subgroup of $G$.

  Show that (ii) fails for infinite subsets $X$.

**1.7**  Let $G$ be a group and let $H$ and $K$ be subgroups of $G$. For a given $x \in G$ define the double coset $HxK$ by

$$HxK = \{hxk \mid h \in H, k \in K\}.$$

If $yK$ is a left coset of $K$, show that either $HxK \cap yK = \emptyset$ or $yK \subseteq HxK$. Hence show that for all $x, y \in G$ either $HxK \cap HyK = \emptyset$ or $HxK = HyK$.

**1.8**  Let $n$ be a prime power and let $C_n$ be a cyclic group of order $n$. If $H$ and $K$ are subgroups of $C_n$, prove that either $H$ is a subgroup of $K$ or $K$ is a subgroup of $H$. Suppose, conversely, that $C_n$ is a cyclic group of order $n$ with the property that, for any two subgroups $H$ and $K$ of $C_n$, either $H$ is a subgroup of $K$ or $K$ is a subgroup of $H$. Is $n$ necessarily a prime power?

**1.9**  Let $G$ be a group. Given a subgroup $H$ of $G$, define

$$H_G = \bigcap_{g \in G} g^{-1} H g.$$

Prove that $H_G$ is a normal subgroup of $G$ and that if $K$ is a subgroup of $H$ that is normal in $G$ then $K$ is a normal subgroup of $H_G$.

  Now let $G = \text{GL}(2, \mathbb{Q})$ and let $H$ be the subgroup of non-singular diagonal matrices. Determine $H_G$. In this case, to what well-known group is $H_G$ isomorphic?

**1.10**  Let $H$ be the subset of $\text{Mat}_{2 \times 2}(\mathbb{C})$ that consists of the elements

$$\begin{bmatrix} 1 & 0 \\ 0 & 1 \end{bmatrix}, \begin{bmatrix} -1 & 0 \\ 0 & -1 \end{bmatrix}, \begin{bmatrix} 0 & 1 \\ -1 & 0 \end{bmatrix}, \begin{bmatrix} 0 & -1 \\ 1 & 0 \end{bmatrix},$$

$$\begin{bmatrix} 0 & i \\ i & 0 \end{bmatrix}, \begin{bmatrix} 0 & -i \\ -i & 0 \end{bmatrix}, \begin{bmatrix} -i & 0 \\ 0 & i \end{bmatrix}, \begin{bmatrix} i & 0 \\ 0 & -i \end{bmatrix}.$$

Prove that $H$ is a non-abelian group under matrix multiplication (called the *quaternion group*). Find all the elements of order 2 in $H$. Find also all the subgroups of $H$. Which of the subgroups are normal? Does $H$ have a quotient group that is isomorphic to the cyclic group of order 4?

**1.11** The *dihedral group* $D_{2n}$ is the subgroup of $GL(2, \mathbb{C})$ that is generated by the matrices

$$\begin{bmatrix} 0 & 1 \\ 1 & 0 \end{bmatrix}, \quad \begin{bmatrix} \alpha & 0 \\ 0 & \alpha^{-1} \end{bmatrix}$$

where $\alpha = e^{2\pi i/n}$.

Prove that $|D_{2n}| = 2n$ and that $D_{2n}$ contains a cyclic subgroup of index 2.

Let $G$ be the subgroup of $GL(2, \mathbb{Z}_n)$ given by

$$G = \left\{ \begin{bmatrix} \epsilon & k \\ 0 & 1 \end{bmatrix} \mid \epsilon = \pm 1, k \in \mathbb{Z}_n \right\}.$$

Prove that $G$ is isomorphic to $D_{2n}$. Show also that, for every positive integer $n$, $D_{2n}$ is a quotient group of the subgroup $D_\infty$ of $GL(2, \mathbb{Z})$ given by

$$D_\infty = \left\{ \begin{bmatrix} \epsilon & k \\ 0 & 1 \end{bmatrix} \mid \epsilon = \pm 1, k \in \mathbb{Z} \right\}.$$

**1.12** Let $\mathbb{Q}^+, \mathbb{R}^+, \mathbb{C}^+$ denote respectively the additive groups of rational, real, complex numbers; and let $\mathbb{Q}^\bullet, \mathbb{R}^\bullet, \mathbb{C}^\bullet$ be the corresponding multiplicative groups. If $U = \{z \in \mathbb{C} \mid |z| = 1\}$ and $\mathbb{Q}^\bullet_{>0}, \mathbb{R}^\bullet_{>0}$ are the multiplicative subgroups of positive rationals and reals, prove that

(i) $\mathbb{C}^+/\mathbb{R}^+ \simeq \mathbb{R}^+$;
(ii) $\mathbb{C}^\bullet/\mathbb{R}^\bullet_{>0} \simeq U$;
(iii) $\mathbb{C}^\bullet/U \simeq \mathbb{R}^\bullet_{>0} \simeq \mathbb{R}^\bullet/C_2$;
(iv) $\mathbb{R}^\bullet/\mathbb{R}^\bullet_{>0} \simeq C_2 \simeq \mathbb{Q}^\bullet/\mathbb{Q}^\bullet_{>0}$;
(v) $\mathbb{Q}^\bullet/C_2 \simeq \mathbb{Q}^\bullet_{>0}$.

**1.13** Let $p$ be a fixed prime. Denote by $\mathbb{Z}_{p^\infty}$ the $p^n$th roots of unity for all positive integers $n$. Then $\mathbb{Z}_{p^\infty}$ is a subgroup of the group of non-zero complex numbers under multiplication.

Prove that every proper subgroup of $\mathbb{Z}_{p^\infty}$ is a finite cyclic group; and that every non-trivial quotient group of $\mathbb{Z}_{p^\infty}$ is isomorphic to $\mathbb{Z}_{p^\infty}$.

Prove that $\mathbb{Z}_{p^\infty}$ and $\mathbb{Q}^+$ satisfy the property that every finite subset generates a cyclic group.

**1.14** Show that if no element of a 2–group $G$ has order 4 then $G$ is abelian.

Show that the dihedral and quaternion groups of order 8 are the only non-abelian groups of order 8. Show further that these two groups are not isomorphic.

## 1: Subgroups

**1.15**  According to Lagrange's theorem, what are the possible orders of sub-groups of $S_4$? For each kind of cycle structure in $S_4$, write down an element with that cycle structure, and determine the total number of such elements. State the order of the elements of each type.

What are the orders of the elements of $S_4$, and how many are there of each order? How many subgroups of order 2 does $S_4$ have, and how many of order 3? Find all the cyclic subgroups of $S_4$ that are of order 4. Find all the non-cyclic subgroups of order 4.

Find all the subgroups of order 6, and all of order 8. Find also a subgroup of order 12.

Find an abelian normal subgroup $V$ of $S_4$. Is $S_4/V$ isomorphic to some subgroup of $S_4$?

Does $A_4$ have a subgroup of order 6?

**1.16**  Consider the subgroup of $S_8$ that is generated by $\{a, b\}$ where

$$a = (1234)(5678) \quad \text{and} \quad b = (1537)(2846).$$

Determine the order of this subgroup and show that it is isomorphic to the quaternion group. Is it isomorphic to any of the subgroups of order 8 in $S_4$?

**1.17**  Suppose that $p$ is a permutation which, when decomposed into a product of disjoint cycles, has all these cycles of the same length. Prove that $p$ is a power of some cycle $\vartheta$.

Prove conversely that if $\vartheta = (1\,2\,\cdots\,m)$ then $\vartheta^s$ decomposes into a product of h.c.f.$(m, s)$ disjoint cycles of length $m/$h.c.f.$(m, s)$.

**1.18**  Let SL$(2, p)$ be the group of $2 \times 2$ matrices of determinant 1 with entries in the field $\mathbb{Z}_p$ (where $p$ is a prime). Show that SL$(2, p)$ contains $p^2(p-1)$ elements of the form

$$\begin{bmatrix} a & b \\ c & d \end{bmatrix}$$

where $a \neq 0$. Show also that SL$(2, p)$ contains $p(p-1)$ elements of the form

$$\begin{bmatrix} 0 & b \\ c & d \end{bmatrix}.$$

Deduce that $|\text{SL}(2, p)| = p(p-1)(p+1)$.

If $Z$ denotes the centre of SL$(2, p)$ define

$$\text{PSL}(2, p) = \text{SL}(2, p)/Z.$$

5

Show that $|\mathrm{PSL}(2,p)| = \frac{1}{2}p(p-1)(p+1)$ if $p \neq 2$.

More generally, consider the group $\mathrm{SL}(n,p)$ of $n \times n$ matrices of determinant 1 with entries in the field $\mathbb{Z}_p$. Using the fact that the rows of a non-singular matrix are linearly independent, prove that

$$|\mathrm{SL}(n,p)| = \frac{1}{p-1} \prod_{i=0}^{n-1} (p^n - p^i).$$

**1.19** Let $F$ be a field in which $1 + 1 \neq 0$ and consider the group $\mathrm{SL}(2,F)$ of $2 \times 2$ matrices of determinant 1 with entries in $F$. Prove that if $A \in \mathrm{SL}(2,F)$ then $A^2 = -I_2$ if and only if $\mathrm{tr}(A) = 0$ (where $\mathrm{tr}(A)$ is the *trace* of $A$, namely the sum of its diagonal elements).

Let $\mathrm{PSL}(2,F)$ be the group $\mathrm{SL}(2,F)/Z(\mathrm{SL}(2,F))$ and denote by $\overline{A}$ the image of $A \in \mathrm{SL}(2,F)$ under the natural morphism $\natural : \mathrm{SL}(2,F) \to \mathrm{PSL}(2,F)$. Show that $\overline{A}$ is of order 2 if and only if $\mathrm{tr}(A) = 0$.

**1.20** Show that $C_2 \times C_2$ is a non-cyclic group of order 4. Prove that if $G$ is a non-cyclic group of order 4 then $G \simeq C_2 \times C_2$.

**1.21** If $p,q$ are primes show that the number of proper non-trivial subgroups of $C_p \times C_q$ is greater than or equal to 2, and that equality holds if and only if $p \neq q$.

**1.22** If $G,H$ are simple groups show that $G \times H$ has exactly two proper non-trivial normal subgroups unless $|G| = |H|$ and is a prime.

**1.23** Is the cartesian product of two periodic groups also periodic? Is the cartesian product of two torsion-free groups also torsion-free?

**1.24** Let $G$ be a group and let $A,B$ be normal subgroups of $G$ such that $G = AB$. If $A \cap B = N$ prove that

$$G/N \simeq A/N \times B/N.$$

Show that this result fails if $G = AB$ where the subgroup $A$ is normal but the subgroup $B$ is not.

**1.25** Let $f : G \to H$ be a group morphism. Suppose that $A$ is a normal subgroup of $G$ and that the restriction of $f$ to $A$ is an isomorphism onto $H$. Prove that

$$G \simeq A \times \mathrm{Ker}\, f.$$

Is this result true without the condition that $A$ be normal?

Deduce that (using the notation defined in question 1.12)

(i) $\mathbb{C}^+ \simeq \mathbb{R}^+ \times \mathbb{R}^+$;

(ii) $\mathbb{Q}^{\bullet} \simeq \mathbb{Q}^{\bullet}_{>0} \times C_2$;

(iii) $\mathbb{R}^{\bullet} \simeq \mathbb{R}^{\bullet}_{>0} \times C_2$;

(iv) $\mathbb{C}^{\bullet} \simeq \mathbb{R}^{\bullet}_{>0} \times U$.

**1.26** Find all the subgroups of $C_2 \times C_2$. Draw the subgroup Hasse diagram.

Prove that if $G$ is a group whose subgroup Hasse diagram is identical to that of $C_2 \times C_2$ then $G \simeq C_2 \times C_2$.

**1.27** Find all the subgroups of $C_2 \times C_2 \times C_2$ and draw the subgroup Hasse diagram.

**1.28** Consider the set of integers $n$ with $1 \leq n \leq 21$ and $n$ coprime to 21. Show that this set forms an abelian group under multiplication modulo 21, and that this group is isomorphic to $C_2 \times C_6$. Is this group cyclic?

Is the set

$$\{n \in \mathbb{Z} \mid 1 \leq n \leq 12, \ n \text{ coprime to } 12\}$$

a cyclic group under multiplication modulo 12?

**1.29** Determine which of the following groups are decomposable into a cartesian product of two non-trivial subgroups :

$$S_4, \quad S_5, \quad A_4, \quad A_5, \quad \mathbb{R}^{\bullet}, \quad C_6, \quad C_8, \quad \mathbb{C}^{+}, \quad \mathbb{Z}_{p^{\infty}}.$$

**1.30** Let $G$ be an abelian group and let $H$ be a subgroup of $G$. Suppose that, given $h \in H$ and $n \in \mathbb{N}$, the equation $x^n = h$ has a solution in $G$ if and only if it has a solution in $H$. Show that given $xH$ there exists $y \in xH$ with $y$ of the same order in $G$ as $xH$ has in $G/H$. Deduce that if $G/H$ is cyclic then there is a subgroup $K$ of $G$ with $G \simeq H \times K$.

**1.31** Let $G$ be an abelian group. If $x, y \in G$ have orders $m, n$ respectively, show that $xy$ has order at most $mn$. Show also that if $z \in G$ has order $mn$ where $m$ and $n$ are coprime then $z = xy$ where $x, y \in G$ satisfy $x^m = y^n = 1$. Deduce that $x$ and $y$ have orders $m, n$ respectively.

Extend this result to the case where $z$ has order $m_1 m_2 \cdots m_k$ where $m_1, \ldots, m_k$ are pairwise coprime.

Hence prove that if $G$ is a finite abelian group of order

$$p_1^{\alpha_1} p_2^{\alpha_2} \cdots p_k^{\alpha_k}$$

where $p_1, \ldots, p_k$ are distinct primes then

$$G = H_1 \times H_2 \times \cdots \times H_k$$

where $H_i = \{x \in G \mid x^{p_i^{\alpha_i}} = 1\}$ for $i = 1, \ldots, k$. Show also that if $r$ divides $|G|$ then $G$ has a subgroup of order $r$.

**1.32** Let $H$ be a subgroup of a group $G$. Prove that the intersection of all the conjugates of $H$ is a normal subgroup of $G$.

If $x \in G$ is it possible that

$$A = \{g^{-1}xg \mid g \in G\}$$

is a subgroup of $G$? Can $A$ be a normal subgroup? Can $A$ be a subgroup that is not normal?

**1.33** Are all subgroups of order 2 conjugate in $S_4$? What about all subgroups of order 3?

Are the elements $(1\,2\,3)$ and $(2\,3\,4)$ conjugate in $A_4$?

**1.34** Show that a subgroup $H$ of a group $G$ is normal if and only if it is a union of conjugacy classes.

Exhibit an element from each conjugacy class of $S_4$ and state how many elements there are in each class. Deduce that the only possible orders for non-trivial proper normal subgroups of $S_4$ are 4 and 12. Show also that normal subgroups of orders 4 and 12 do exist in $S_4$.

**1.35** Exhibit an element from each conjugacy class of $S_5$. How many elements are there in each conjugacy class? What are the orders of the elements of $S_5$? Find all the non-trivial proper normal subgroups of $S_5$.

Find the conjugacy classes of $A_5$ and deduce that it has no proper non-trivial normal subgroups.

**1.36** If $G$ is a group and $a \in G$ prove that the number of elements in the conjugacy class of $a$ is the index of $\mathcal{N}_G(a)$ in $G$. Deduce that in $S_n$ the only elements that commute with a cycle of length $n$ are the powers of that cycle.

Suppose that $n$ is an odd integer, with $n \geq 3$. Prove that there are two conjugacy classes of cycles of length $n$ in $A_n$. Show also that each of these classes contains $\frac{1}{2}(n-1)!$ elements.

Show that if $n$ is an even integer with $n \geq 4$ then there are two conjugacy classes of cycles of length $n-1$ in $S_n$, and that each of these classes contains $\frac{1}{2}n(n-2)!$ elements.

**1.37** If $G$ is a group and $a \in G$ prove that the conjugacy class containing $a$ and that containing $a^{-1}$ have the same number of elements.

Suppose now that $|G|$ is even. Show that there is at least one $a \in G$ with $a \neq 1$ such that $a$ is conjugate to $a^{-1}$.

**1.38** Find the conjugacy classes of the dihedral group $D_{2n}$ when $n$ is odd. What are the classes when $n$ is even?

**1.39** Let $G$ be a group and let $H$ and $K$ be conjugate subgroups of $G$. Prove that $N_G(H)$ and $N_G(K)$ are conjugate.

**1.40** Let $H$ be a normal subgroup of a group $G$ with $|H| = 2$. Prove that $H \subseteq Z(G)$.

Is it necessarily true that $H \subseteq G'$?

Prove that if $G$ contains exactly one element $x$ of order 2 then $\langle x \rangle \subseteq Z(G)$.

**1.41** Suppose that $N$ is a normal subgroup of a group $G$ with the property that $N \cap G' = 1$. Prove that $N \subseteq Z(G)$ and deduce that

$$Z(G/N) = Z(G)/N.$$

# 2: Automorphisms and Sylow theory

An isomorphism $f : G \to G$ is called an automorphism on $G$. The automorphisms on a group $G$ form, under composition of mappings, a group $\operatorname{Aut} G$. Conjugation by a fixed element $g$ of $G$, namely the mapping $\varphi_g : G \to G$ described by $x \mapsto \varphi_g(x) = g^{-1}xg$, is an automorphism on $G$. The inner automorphism group $\operatorname{Inn} G = \{\varphi_g \mid g \in G\}$ is a normal subgroup of $\operatorname{Aut} G$, and the quotient group $\operatorname{Aut} G/\operatorname{Inn} G$ is called the outer automorphism group of $G$. For example, the cyclic group $C_n$ (being abelian) has trivial inner automorphism group, and $\vartheta : C_n \to C_n$ given by $\vartheta(g) = g^{-1}$ is an (outer) automorphism of order 2. A subgroup $H$ of a group $G$ is normal if and only if $\vartheta(H) \subseteq H$ for every $\vartheta \in \operatorname{Inn} G$, and is called characteristic if $\vartheta(H) \subseteq H$ for every $\vartheta \in \operatorname{Aut} G$.

For finite groups, the converse of Lagrange's theorem is false. However, a partial converse is provided by the important theorems of Sylow. A group $P$ is called a $p$–group if every element has order a power of $p$ for a fixed prime $p$. In this case, if $P$ is finite, $|P|$ is also a power of $p$. If $G$ is a group with $|G| = p^n k$ where $k$ is coprime to $p$ then a subgroup of order $p^n$ is called a Sylow $p$–subgroup. In this situation we have the following results, with which we assume the reader is familiar :

(a) $G$ has a subgroup of order $p^m$ for every $m \leq n$;
(b) every $p$–subgroup of $G$ is contained in a Sylow $p$–subgroup;
(c) any two Sylow $p$–subgroups are conjugate in $G$;
(d) the number of Sylow $p$–subgroups of $G$ is congruent to 1 modulo $p$ and divides $|G|$.

**2.1**  Let $p$ be a prime. Use the class equation to show that every finite $p$–group has a non-trivial centre. Deduce that all groups of order $p^2$ are abelian.

# 2: Automorphisms and Sylow theory

List all the groups of order 9.

**2.2** Let $G$ be a group and let $\vartheta \in \operatorname{Aut} G$. If $A$ and $B$ are subgroups of $G$ prove that $\vartheta(A \cap B)$ is a subgroup of $\vartheta A \cap \vartheta B$. Is it necessarily true that $\vartheta(A \cap B) = \vartheta A \cap \vartheta B$?

**2.3** Let $G$ be a group and let $\operatorname{Inn} G$ be the group of inner automorphisms on $G$. Prove that $\operatorname{Inn} G$ is a normal subgroup of $\operatorname{Aut} G$ and that

$$\operatorname{Inn} G \simeq G/Z(G).$$

The two non-abelian groups of order 8 are the dihedral group $D_8$ with presentation

$$D_8 = \langle\, a, b \mid a^2 = 1,\ b^4 = 1,\ a^{-1}ba = b^{-1} \,\rangle$$

and the quaternion group $Q_8$ with presentation

$$Q_8 = \langle\, x, y \mid x^4 = 1,\ x^2 = y^2,\ y^{-1}xy = x^{-1} \,\rangle.$$

Show that $Z(D_8) = \langle b^2 \rangle \simeq C_2$ and $Z(Q_8) = \langle x^2 \rangle \simeq C_2$. Deduce that $\operatorname{Inn} D_8 \simeq \operatorname{Inn} Q_8$.

**2.4** Let $G$ be a group with the property that it cannot be decomposed into the direct product of two non-trivial subgroups. Does every subgroup of $G$ have this property? Does every quotient group of $G$ have this property?

**2.5** If $G$ is a group such that $G/Z(G)$ is cyclic, prove that $G$ is abelian. Deduce that a group with a cyclic automorphism group is necessarily abelian.

**2.6** Find the automorphism group of the symmetric group $S_3$.

**2.7** Prove that $C_2 \times C_2$ and $S_3$ have isomorphic automorphism groups.

**2.8** Let $G$ be a group with $Z(G) = \{1\}$. Prove that $Z(\operatorname{Aut} G) = \{1\}$. Is the converse true in general?

**2.9** Let $\mathbb{Z}_p$ denote the field of integers modulo $p$ where $p$ is a prime, and let $\mathbb{Z}_p^n$ be an $n$–dimensional vector space over $\mathbb{Z}_p$. Prove that the additive group of $\mathbb{Z}_p^n$ is isomorphic to the group

$$C_p \times C_p \times \cdots \times C_p$$

consisting of $n$ copies of $C_p$.

Show that every element of $\operatorname{Aut} G$ corresponds to an invertible linear transformation on $\mathbb{Z}_p^n$.

Deduce that

$$\operatorname{Aut} G \simeq \operatorname{GL}(n, p).$$

**2.10** Find all groups $G$ with $\operatorname{Aut} G = \{1\}$.

11

**2.11**  A subgroup $H$ of a group $G$ is called *fully invariant* if $\vartheta(H) \subseteq H$ for every group morphism $\vartheta : G \to G$. Which of the following statements are true?

(a) The derived group of a group is fully invariant.

(b) The centre of a group is fully invariant.

(c) $A_4$ contains a normal subgroup that is not fully invariant.

(d) $G^n = \langle g^n \mid g \in G \rangle$ is a fully invariant subgroup of $G$.

(e) $G_n = \langle g \in G \mid g^n = 1 \rangle$ is a fully invariant subgroup of $G$.

**2.12**  Let $G$ be a group and $C$ a conjugacy class in $G$. If $\alpha \in \operatorname{Aut} G$ prove that $\alpha(C)$ is also a conjugacy class of $G$.

Let $K$ be the set of conjugacy classes of $G$ and define

$$N = \{\alpha \in \operatorname{Aut} G \mid (\forall C \in K)\, \alpha(C) = C\}.$$

Prove that $N$ is a normal subgroup of $\operatorname{Aut} G$.

**2.13**  Let $G$ be a group and $N$ a normal subgroup of $G$. Let $A = \operatorname{Aut} N$ and $I = \operatorname{Inn} N$. If $C$ is the centraliser of $N$ in $G$ prove that $NC$ is a normal subgroup of $G$ and that $G/NC$ is isomorphic to a subgroup of the outer automorphism group $A/I$ of $N$. Show also that $NC/C \simeq I$.

Prove that if the outer automorphism group of $N$ is trivial and $Z(N) = \{1\}$ then $G = N \times C$. Deduce that a group $G$ contains $S_3$ as a normal subgroup if and only if $G = S_3 \times C$ for some normal subgroup $C$ of $G$.

**2.14**  Prove that if $G$ is a group then

(a) a subgroup $H$ is characteristic in $G$ if and only if $\vartheta(H) = H$ for every $\vartheta \in \operatorname{Aut} G$;

(b) the intersection of a family of characteristic subgroups of $G$ is a characteristic subgroup of $G$;

(c) if $H, K$ are characteristic subgroups of $G$ then so is $HK$;

(d) if $H, K$ are characteristic subgroups of $G$ then so is $[H, K]$;

(e) if $H$ is a normal subgroup of $G$, and $K$ is a characteristic subgroup of $H$, then $K$ is a normal subgroup of $G$.

**2.15**  Suppose that $G$ is a finite group and that $H$ is a normal subgroup of $G$ such that $|H|$ is coprime to $|G : H|$. Prove that $H$ is characteristic in $G$.

**2.16**  Let $G$ be a group and let $F$ be the subset consisting of those elements $x$ of $G$ that have only finitely many conjugates in $G$. Prove that $F$ is a subgroup of $G$. Is $F$ a normal subgroup? Is $F$ characteristic in $G$?

**2.17**  If $t \in \mathbb{Q}\backslash\{0\}$ prove that $\vartheta_t : \mathbb{Q}^+ \to \mathbb{Q}^+$ given by $\vartheta_t(r) = tr$ is an automorphism of the (additive) group $\mathbb{Q}^+$. Deduce that the only characteristic subgroups of $\mathbb{Q}^+$ are $\{1\}$ and $\mathbb{Q}^+$.

## 2: *Automorphisms and Sylow theory*

**2.18** Suppose that $G$ is a finite group and that $H$ is a subgroup of $G$. Show that every Sylow $p$-subgroup of $H$ is contained in a Sylow $p$-subgroup of $G$. Prove also that no pair of distinct $p$-subgroups of $H$ can lie in the same Sylow $p$-subgroup of $G$.

Now suppose that that $H$ is normal in $G$ and that $P$ is a Sylow $p$-subgroup of $G$. Prove that $H \cap P$ is a Sylow $p$-subgroup of $H$ and that $HP/H$ is a Sylow $p$-subgroup of $G/H$. Is $H \cap P$ a Sylow $p$-subgroup of $H$ if we drop the condition that $H$ be normal in $G$?

**2.19** Prove that a normal $p$-subgroup of a finite group $G$ is contained in every Sylow $p$-subgroup of $G$.

Suppose that, for every prime $p$ dividing $|G|$, $G$ has a normal Sylow $p$-subgroup. Prove that $G$ is the direct product of its Sylow $p$-subgroups.

**2.20** Determine the structure of the Sylow $p$-subgroups of $A_5$ and find the number of Sylow $p$-subgroups for each prime $p$.

**2.21** Let $G$ be a finite group and let $K$ be a normal subgroup of $G$. Suppose that $P$ is a Sylow $p$-subgroup of $K$. Show that, for all $g \in G$, $g^{-1}Pg$ is also a Sylow $p$-subgroup of $K$. Use the fact that these Sylow $p$-subgroups are conjugate in $K$ to deduce that $G = \mathcal{N}(P)K$. Deduce further that if $\overline{P}$ is a Sylow $p$-subgroup of $G$ and $\mathcal{N}(\overline{P}) \leq H \leq G$ then $\mathcal{N}(H) = H$.

**2.22** Let $G$ be a finite group with the property that all its Sylow subgroups are cyclic. Show that every subgroup of $G$ has this property.

Prove that any two $p$-subgroups of $G$ of the same order are conjugate.

Let $H$ and $N$ be subgroups of $G$ with $N$ normal in $G$. Show that

$$|N \cap H| = \text{h.c.f.}(|N|, |H|),$$
$$|HN| = \text{l.c.m.}(|N|, |H|).$$

Deduce that every normal subgroup of $G$ is characteristic.

**2.23** Use the Sylow theorems to prove that
   (a) every group of order 200 has a normal Sylow 5-subgroup;
   (b) there is no simple group of order 40;
   (c) there is no simple group of order 56;
   (d) every group of order 35 is cyclic.

**2.24** Use the Sylow theorems to prove that
   (a) every group of order 85 is cyclic;
   (b) if $p, q$ are distinct primes then a group of order $p^2 q$ cannot be simple.

**2.25** Let $G$ be a group of order $pq$ where $p, q$ are distinct primes such that $q \not\equiv 1$ modulo $p$. Prove that $G$ has a normal Sylow $p$–subgroup. Show that this result fails if $q \equiv 1$ modulo $p$. Show that if $|G| = pq$ where $p, q$ are distinct primes then $G$ is not simple. Deduce further that if $p, q$ are distinct primes with $p \not\equiv 1$ modulo $q$ and $q \not\equiv 1$ modulo $p$ then every group of order $pq$ is cyclic.

**2.26** Suppose that a group $G$ has the property that if $n$ divides $|G|$ then $G$ has a subgroup of order $n$. Does every subgroup of $G$ have this property?

**2.27** Let $G$ be a finite group and $P$ a Sylow $p$–subgroup of $G$. Suppose that $x, y \in Z(P)$ and are conjugate in $G$. Show that $x, y$ are conjugate in $\mathcal{N}(P)$.

**2.28** Let $G$ be a group with a subgroup $H$ of index $n$ in $G$. Show that there is a largest normal subgroup $K$ of $G$ that is contained in $H$ and that $G/K$ is isomorphic to a subgroup of $S_n$.

    Deduce that if $G$ is a simple group with $|G| = 60$ (there is exactly one such group, namely $A_5$, but this fact is not required) then $G$ has no subgroups of order 15, 20 or 30.

**2.29** Let $G$ be a simple group with $|G| = 168$. Prove that $G$ has eight Sylow 7–subgroups. Show also that if $P$ is a Sylow 7–subgroup of $G$ then $|\mathcal{N}_G(P)| = 21$. Deduce that $G$ contains no subgroup of order 14.

# 3: Series

Given subgroups $A, B$ of a group $G$ we obtain the subgroup

$$[A, B] = \langle [a, b] \mid a \in A, b \in B \rangle.$$

In particular, $[G, G]$ is the derived group of $G$. We define the derived series of $G$ to be the most rapidly descending series with abelian quotients (factors), namely

$$G^{(0)} = G, \qquad (\forall i \geq 1) \quad G^{(i)} = [G^{(i-1)}, G^{(i-1)}].$$

We say that $G$ is soluble of derived length $n$ if $n$ is the least integer with $G^{(n)} = \{1\}$.

Similarly, the most rapidly descending central series and the most rapidly ascending central series of $G$ are the lower central series and the upper central series, defined by

$$\Gamma_1(G) = G, \qquad (\forall i \geq 1) \quad \Gamma_{i+1}(G) = [\Gamma_i(G), G],$$

$$Z_0 = \{1\}, \qquad (\forall i \geq 1) \quad Z_i/Z_{i-1} = Z(G/Z_{i-1})$$

respectively. The lower central series reaches $\{1\}$ in a finite number of steps if and only if the upper central series reaches $G$ in a finite number of steps. In this case $G$ is said to be nilpotent, and the number of factors in either series is the class of $G$.

Every subgroup $H$ of a nilpotent group $G$ is subnormal, in the sense that there is a series

$$H = H_0 \lhd H_1 \lhd \cdots \lhd H_r = G.$$

The final type of series with which we assume the reader is familiar is called a composition series. This is a subnormal series from $\{1\}$ into which no further terms can be properly inserted.

**3.1** Let $G$ be a group. Establish each of the following results concerning commutators.

(a) If $S \leq G$ and $T \leq G$ then $[S, T] = [T, S]$.

(b) If $H \triangleleft G$ and $K \triangleleft G$ then $[H, K] \leq H \cap K$. What does this imply when $H \cap K = \{1\}$?

(c) If $x, y, z \in G$ then

$$[xy, z] = y^{-1}[x, z]y[y, z].$$

Deduce that if $H, K, L$ are normal subgroups of $G$ then

$$[HL, K] = [H, K][L, K].$$

(d) Define $[a, b, c,] = [[a, b], c]$. Prove that

$$[a, bc] = [a, c][a, b][a, b, c]$$

and that

$$[ab, c] = [a, c][a, c, b][b, c].$$

**3.2** Find the upper and lower central series of $G = Q_8 \times C_2$ and show that they do not coincide. Show, however, that the upper and lower central series of $Q_8$ do coincide.

**3.3** Prove that if $G$ is generated by its subnormal abelian subgroups then any quotient group of $G$ is generated by its subnormal abelian subgroups.

Show that every subgroup of a nilpotent group is subnormal. Deduce that a nilpotent group is generated by its subnormal abelian subgroups.

**3.4** Let $A, B, C$ be subgroups of a group $G$ with $B \triangleleft A$. Prove that

$$\frac{A \cap C}{B \cap C} \simeq \frac{B(A \cap C)}{B}.$$

If, in addition, $C \triangleleft G$ prove that

$$\frac{AC}{BC} \simeq \frac{A}{B(A \cap C)}.$$

Use the above results to show that if $H$ is a soluble group then every subgroup and every quotient group of $H$ is soluble.

Prove that if $K$ is a group with $H \triangleleft K$ and both $H$ and $K/H$ are soluble then $K$ is soluble.

Let $G$ be a group with normal subgroups $A$ and $B$ such that $G/A$ and $G/B$ are soluble. Show that $A/(A \cap B)$ is soluble and deduce that so also is $G/(A \cap B)$.

**3.5**  Which of the following statements are true? Give a proof for those that are true and a counter-example to those that are false.

(a) Let $G$ be a group and let $H, K$ be normal soluble subgroups of $G$. Then $HK$ is a normal soluble subgroup of $G$.

(b) Let $G$ be a group and $H, K$ normal abelian subgroups of $G$. Then $HK$ is a normal abelian subgroup of $G$.

(c) Let $G$ be a group and $H, K$ normal $p$–subgroups of $G$. Then $HK$ is a normal $p$–subgroup of $G$.

**3.6**  Let $G$ be a non-trivial finite nilpotent group. Use induction on $|G|$ to prove that every proper subgroup of $G$ is properly contained in its normaliser. Deduce that every Sylow subgroup of $G$ is normal. [*Hint.* Use question 2.21.]

**3.7**  Suppose that $G$ is a group with the properties

(a) $G$ is nilpotent of class 3;

(b) $|G| = 16$.

Prove that $G$ contains a unique cyclic subgroup of order 8.

Give an example of such a group.

**3.8**  A group $G$ is said to be *residually nilpotent* if it has a series of subgroups

$$G = H_1 \geq H_2 \geq \cdots \geq H_i \geq \cdots$$

with $[H_i, G] \leq H_{i+1}$ and $\bigcap_{i=1}^{\infty} H_i = \{1\}$.

Show that a finite group is residually nilpotent if and only if it is nilpotent. Give an example of a residually nilpotent group that is not nilpotent.

Prove that every subgroup of a residually nilpotent group is also residually nilpotent. Show that a quotient group of a residually nilpotent group need not be residually nilpotent.

**3.9**  Establish the identity

$$[xy, z] = y^{-1}[x, z]y[y, z].$$

Hence show that if $A$ is a subgroup of a group $G$ then $[G, A]$ is normal in $G$.

Prove that if $G$ is a group with a non-trivial subgroup $A$ such that $A = [A, G]$ then $G$ cannot be nilpotent.

A minimal normal subgroup of a group is a non-trivial normal subgroup which properly contains no non-trivial normal subgroup of the group. Deduce from the above that every minimal normal subgroup of a nilpotent group is contained in the centre of the group.

**3.10**  Let $p$ be a prime. Prove that every finite $p$–group is nilpotent.

    Let $G = H \times K$ where $|H| = p^2$ and $|K| = p^3$. Prove that if $G$ is non-abelian then $G$ is nilpotent of class 2 and $|Z(G)| = p^3$.

**3.11**  Let $G$ be the multiplicative group

$$\left\{ \begin{bmatrix} 1 & a & b \\ 0 & 1 & c \\ 0 & 0 & 1 \end{bmatrix} \;\middle|\; a, b, c \in \mathbb{Z} \right\}.$$

Find the centre of $G$ and the derived group of $G$. Prove that $G$ is nilpotent and that the upper and lower central series for $G$ coincide.

    Let $t_{12}, t_{13}, t_{23}$ denote the matrices

$$\begin{bmatrix} 1 & 1 & 0 \\ 0 & 1 & 0 \\ 0 & 0 & 1 \end{bmatrix}, \quad \begin{bmatrix} 1 & 0 & 1 \\ 0 & 1 & 0 \\ 0 & 0 & 1 \end{bmatrix}, \quad \begin{bmatrix} 1 & 0 & 0 \\ 0 & 1 & 1 \\ 0 & 0 & 1 \end{bmatrix}$$

respectively. Prove that

$$G = \langle t_{12}, t_{13}, t_{23} \rangle.$$

Find a subnormal series for each of the subgroups

$$\langle t_{12} \rangle, \quad \langle t_{13} \rangle, \quad \langle t_{23} \rangle.$$

**3.12**  Let $X, Y, Z$ be subgroups of a group $G$ and let

$$A = [X, Y, Z], \quad B = [Y, Z, X], \quad C = [Z, X, Y].$$

Prove that if $N$ is a normal subgroup of $G$ that contains two of $A, B, C$ then $N$ contains the third.

[*Hint.* Use the identity $[x, y^{-1}, z]^y [y, z^{-1}, x]^z [z, x^{-1}, y]^x = 1$.]

    Deduce that if $G$ has subgroups $H$ and $K$ such that

$$H = H_0 \geq H_1 \geq H_2 \geq \cdots$$

is a series of normal subgroups of $H$ with $[H_i, K] \leq H_{i+1}$ for all $i \geq 0$ then $[H_i, \Gamma_n(K)] \leq H_{i+n}$ for every $n \in \mathbb{N}$ where $\Gamma_n(K)$ is the $n$th term of the lower central series for $K$.

    Suppose that $G$ has lower central series $G = \Gamma_1 \geq \Gamma_2 \geq \cdots$, upper central series $\{1\} = Z_0 \leq Z_1 \leq \cdots$, and derived series $G = G^{(0)} \geq G^{(1)} \geq \cdots$. Prove that

(a) $[\Gamma_m, \Gamma_n] \leq \Gamma_{m+n}$;
(b) $[Z_m, \Gamma_n] \leq Z_{m-n}$;
(c) $[Z_m, \Gamma_m] = \{1\}$;
(d) $G^{(r)} \leq \Gamma_{2^r}$;
(e) if $G = G^{(1)}$ then $Z_1 = Z_2$.

# 3: Series

**3.13** Let $G$ be a group with $|G/Z(G)| = p^n$. Let $x \in Z_2(G)$ and $N = \langle [x, g] \mid g \in G \rangle$. Prove that $|N| < p^n$.

Use induction to prove that $G'$ is a $p$–group of order at most $p^{\frac{1}{2}n(n-1)}$.

**3.14** Let $G$ be a finite group and let $\Phi$ be the intersection of all the maximal subgroups of $G$. Prove that if $H$ is a subgroup of $G$ such that $G = \Phi H$ then $H = G$.

Let $T$ be a Sylow $p$–subgroup of $\Phi$ and let $g \in G$. By considering $T$ and $T^g$, prove that $g \in N_G(T)\Phi$. Deduce that every Sylow $p$–subgroup of $\Phi$ is normal.

**3.15** A group $G$ is said to *satisfy the maximum condition for subgroups* if for every chain of subgroups

$$H_1 \leq H_2 \leq \cdots \leq H_n \leq \cdots$$

there is an integer $N$ such that $(\forall m \geq N)\ H_m = H_N$.

Prove that $G$ satisfies the maximum condition for subgroups if and only if every subgroup of $G$ is finitely generated.

A group $G$ is called *polycyclic* if $G$ has a series

$$G = H_0 \geq H_1 \geq \cdots \geq H_r = \{1\}$$

with $H_i \triangleleft H_{i-1}$ and $H_{i-1}/H_i$ cyclic for $i = 1, \ldots, r$.

Prove that a group $G$ is polycyclic if and only if $G$ is soluble and satisfies the maximum condition for subgroups.

**3.16** Suppose that $A$ and $B$ are abelian subgroups of a group $G$ such that $G = AB$. Use the relation

$$[xy, z] = y^{-1}[x, z]y[y, z]$$

to prove that $[A, B]$ is normal in $G$ and hence show that $G' = [A, B]$.

Prove further that if $a_1, a_2 \in A$ and $b_1, b_2 \in B$ then

$$(a_2 b_2)^{-1}[a_1, b_1]a_2 b_2 = (b_2 a_2)^{-1}[a_1, b_1]b_2 a_2$$

and deduce that $[A, B]$ is abelian.

Conclude that $G$ is soluble of derived length at most 2.

**3.17** Let $G$ be a finite nilpotent group. Prove that every maximal subgroup of $G$ is normal.

Let $S$ be a Sylow $p$–subgroup of $G$ and suppose that $N_G(S)$ is properly contained in $G$. Let $M$ be a maximal subgroup of $G$ containing $N_G(S)$. Show that if $g \in G$ then $S$ and $g^{-1}Sg$ are Sylow $p$–subgroups of $M$ and derive the contradiction that $g \in M$. Hence deduce that $N_G(S) = G$ and that $G$ has just one Sylow $p$–subgroup for each prime $p$ that divides the order of $G$.

Prove also that $G$ is the direct product of its Sylow subgroups.

**3.18** Let $M$ be a maximal subgroup of a finite soluble group $G$. Let $K$ be the intersection of all the subgroups of $G$ that are conjugate to $M$. Prove that $K$ is the largest normal subgroup of $G$ contained in $M$.

Let $H/K$ be a minimal normal subgroup of $G/K$. Prove that $G = HM$ and that $H \cap M = K$.

Deduce that the index of $M$ in $G$ is equal to the order of $H/K$.

**3.19** A group $G$ is called *metacyclic* if it has a normal subgroup $N$ such that $N$ and $G/N$ are cyclic.

Prove that every subgroup and every quotient group of a metacyclic group is also metacyclic.

Show that the group described by

$$\langle\, a, b \mid a^2 = 1,\ b^8 = 1,\ aba = b^7 \,\rangle$$

is metacyclic.

**3.20** Let $F$ be a field. If $a \in F$ and $1 \le i < j \le n$ let $t_{ij}(a)$ be the matrix in $\mathrm{GL}(n, F)$ that differs from the identity in having $a$ in the $(i,j)$–th position. Let

$$T_n(F) = \langle\, t_{ij}(a) \mid 1 \le i < j \le n,\ a \in F \,\rangle.$$

Prove that $T_n(F)$ is the subgroup of $\mathrm{GL}(n, F)$ consisting of all upper triangular $n \times n$ matrices over $F$ of the form

$$\begin{bmatrix} 1 & \star & \star & \cdots & \star \\ 0 & 1 & \star & \cdots & \star \\ 0 & 0 & 1 & \cdots & \star \\ \vdots & \vdots & \vdots & \ddots & \vdots \\ 0 & 0 & 0 & \cdots & 1 \end{bmatrix}.$$

For $1 \le k \le n - 1$ let $H_k = \langle\, t_{ij}(a) \mid j - i \ge k,\ a \in F \,\rangle$. Prove that

$$T_n(F) = H_1 \ge H_2 \ge \cdots \ge H_{n-1} \ge \{1\}$$

is a central series for $T_n(F)$.

If $F = \mathbb{Z}_p$ for some prime $p$, show that $T_n(F)$ is a Sylow $p$–subgroup of $\mathrm{SL}(n, F)$.

**3.21** Show that the groups $S_3, S_4$ and $S_5$ have unique composition series. Find the composition series for each of these groups.

What can you say about a composition series of $S_n$ when $n \ge 5$?

**3.22** Let $G$ be a group of order $p^r q^s$ where $p$ and $q$ are distinct primes. Suppose that $G$ has composition series

$$G = A_1 > A_2 > \cdots > A_{r+s+1} = \{1\},$$
$$G = B_1 > B_2 > \cdots > B_{r+s+1} = \{1\}$$

such that $|A_{r+1}| = q^s$ and $|B_{s+1}| = p^r$. Show that $A_{r+1}$ and $B_{s+1}$ are normal subgroups of $G$ and deduce that $G$ is the direct product of these subgroups.

# 4: Presentations

Given an abelian group $G$ with a presentation

$$\langle\, x_1, \ldots, x_n \mid r_1 = 1,\ r_2 = 1,\ \ldots,\ r_m = 1 \,\rangle$$

the relation matrix of the presentation is the $m \times n$ matrix $A = [a_{ij}]$ where $a_{ij}$ is the exponent sum of $x_j$ in the relation $r_i = 1$. Now $A$ can be reduced by elementary row and column operations, in which only integer multiples are used, to a diagonal matrix $D = \mathrm{diag}\{d_1, \ldots, d_t\}$ where $t = \min\{n, m\}$. This is equivalent to finding invertible integer matrices $P, Q$ such that $PAQ^{-1} = D$. We can assume that $d_1, \ldots, d_k$ are non-zero and $d_{k+1}, \ldots, d_t$ are zero. Then if $C$ is the direct product of $n - k$ copies of $C_\infty$ we have

$$G \simeq C_{d_1} \times C_{d_2} \times \cdots \times C_{d_k} \times C.$$

If $G$ is a group then $G/G'$ is abelian, and $G$ is called perfect if $G/G'$ is the trivial group. If $G$ is given by the presentation $\langle\, X \mid R \,\rangle$ then $G/G'$ is given by the presentation $\langle\, X \mid R, C \,\rangle$ where

$$C = \{[x_i, x_j] \mid x_i, x_j \in X\}.$$

This is a special case of von Dyck's theorem which shows that adding relations to a group presentation leads to a quotient group. In fact, von Dyck's theorem lies behind the method of showing that a given presentation defines a particular group.

**4.1**  Let $G$ be the abelian group

$$\langle\, a, b, c \mid a^{37}b^{27}c^{47} = a^{52}b^{37}c^{67} = a^{59}b^{44}c^{74} = 1,$$
$$ab = ba,\ bc = cb,\ ca = ac \,\rangle.$$

Express $G$ as a direct product of cyclic groups.

By adding the relation $a^3b^2c^4 = 1$ to those of $G$ show that $a^3b^2c^4$ is not the identity of $G$. Deduce that $a^3b^2c^4$ is an element of order 5 in $G$.

Find an element of order 7 and an element of order 35 in $G$.

**4.2** Express each of the following as a direct product of cyclic groups :

(a) $\langle a, b, c \mid a^2b^3c^6 = a^4b^9c^4 = 1, \ ab = ba, \ bc = cb, \ ca = ac \rangle$;
(b) $\langle a, b, c \mid a^2b^3c^6 = a^4b^9c^4 = a^3b^3c^2 = 1, \ ab = ba, \ bc = cb, \ ac = ca \rangle$.

**4.3** Let $G$ be the group with presentation

$$G = \langle a, b \mid abab^2 = 1 \rangle.$$

Show that $G$ is abelian. Describe the structure of $G$.

**4.4** Let $G$ be an abelian group. Show that the elements of finite order form a subgroup $T$. Let $Q$ consist of the elements of infinite order together with the identity element. Find a necessary and sufficient condition for $Q$ to be a subgroup of $G$.

Let $\Pi$ be the set of primes and define a group $G$ as follows. Let $X = \bigcup_{p \in \Pi} \mathbb{Z}_p$ and let $G$ be the set of mappings $f : \Pi \to X$ such that $f(p) \in \mathbb{Z}_p$ for every $p \in \Pi$. Given $f, g \in G$ define $f + g : \Pi \to X$ to be the mapping given by

$$(\forall p \in \Pi) \qquad (f + g)(p) = f(p) + g(p).$$

Show that $G$ is an abelian group containing elements of prime order for every prime, and elements of infinite order.

Prove that the subgroup $T$ in this case consists of those mappings $f \in G$ with the property that there are only finitely many $p \in \Pi$ with $f(p) \neq 0$.

**4.5** Let $G$ be the group with presentation

$$G = \langle a, b, c \mid a^n b^m c^m = a^m b^n c^m = a^m b^m c^n = 1 \rangle.$$

Prove that $G/G'$ is infinite if and only if $m = n$ or $2m = -n$.

Show further that $G$ is perfect if and only if $G$ is the trivial group.

**4.6** If $n$ is an integer that is coprime to 6 show that there is an integer $k$ such that the group

$$\langle x, y \mid x^2 = (xy)^3, \ (xy^4xy^{\frac{1}{2}(n+1)})^2 y^n x^{2k} = 1 \rangle$$

is perfect.

**4.7** Find $G/G'$ when $G$ is given by

$$\langle\, a, b \ \mid \ a^n = 1, \ b^2 = (ab)^3, \ (a^{\frac{1}{2}(n+1)}ba^4b)^2 = 1 \,\rangle,$$

where $n$ is an odd integer.

**4.8** Let $G$ be the group generated by $R, S, T, U, V, W, X$ subject to the relations

$$R^x = S^a T^b U^c,$$
$$SVR^y = 1,$$
$$V^y T^a U^d = 1,$$
$$T^{-1}WV^z = 1,$$
$$W^{-z}U^a = 1,$$
$$UX^{-1}W^t = 1,$$
$$X^t = 1.$$

Find a presentation for $G$ on the generators $R, S, T, U$.

Show that whether $G/G'$ is finite depends only on $x, y, z, t, a$. Give precise conditions for $G/G'$ to be finite. Find values of $a, b, c, d, x, y, z, t$ so that

(a) $G$ is perfect;
(b) $G/G' \simeq C_{16}$;
(c) $G/G' \simeq C_2 \times C_4 \times C_8$.

**4.9** Let $G$ be the group with presentation

$$\langle\, a_1, \ldots, a_{2m} \ \mid \ a_i = a_{i+2}a_{i+m+1}, \ a_{i+2} = a_{i+1}a_i, \ (i = 1, \ldots, 2m) \,\rangle$$

where the subscripts are reduced modulo $2m$ to lie between $1$ and $2m$.

Prove that $G = \langle\, a_1, a_2 \,\rangle$.

Define the Fibonacci sequence by

$$f_1 = 1, \ f_2 = 1, \ (\forall n \geq 1) \ f_{n+2} = f_{n+1} + f_n.$$

Prove that

$$(\forall n \geq 2) \qquad f_{n-1}f_{n+1} - f_n^2 = (-1)^n.$$

If $g_n = f_{n-1} + f_{n+1}$ show that

$$|G/G'| = \begin{cases} 2 + g_m & \text{if } m \text{ is even;} \\ g_m & \text{if } m \text{ is odd.} \end{cases}$$

# 4: Presentations

**4.10**  Let $G$ be defined by

$$G = \langle\, a, b \mid a^{2^{n-1}} = 1,\ bab^{-1} = a^{-1},\ b^2 = a^{2^{n-2}} \,\rangle$$

and let $H$ be defined by

$$H = \langle\, a, b \mid a^{2^{n-2}} = b^2 = (ab)^2 \,\rangle.$$

Prove that $G \simeq H$ by showing that the relations of $G$ imply those of $H$ and conversely.

**4.11**  Show that the groups

$$G = \langle\, a, b \mid a^4 = 1,\ a^2 = b^2,\ ab = ba^3 \,\rangle$$
$$H = \langle\, a, b \mid a = bab,\ b = aba \,\rangle$$
$$K = \langle\, a, b \mid ab = c,\ bc = a,\ ca = b \,\rangle$$

are isomorphic.

Show that

$$a = \begin{bmatrix} 0 & 1 \\ -1 & 0 \end{bmatrix}, \quad b = \begin{bmatrix} 0 & i \\ i & 0 \end{bmatrix}, \quad c = \begin{bmatrix} i & 0 \\ 0 & -i \end{bmatrix}$$

satisfy the above presentations, and that the multiplicative group generated by $a$ and $b$ is the quaternion group. Deduce that all of the above presentations are presentations of the quaternion group.

**4.12**  Let $G$ be a group and suppose that

$$G = \langle\, a_1, \ldots, a_n \,\rangle.$$

If $a_n \in G' \cap Z(G)$, prove that

$$G = \langle\, a_1, \ldots, a_{n-1} \,\rangle.$$

Deduce that if $H/A \simeq Q_8$ with $A \leq Z(H) \cap H'$ then $H \simeq Q_8$.

**4.13**  Let $G$ be the group with presentation

$$G = \langle\, x, y \mid x^2 y = y^2 x,\ x^8 = 1 \,\rangle.$$

Prove that $(xy)^4 = 1$ and $y^8 = 1$.

**4.14**   Let $G$ be the group with presentation

$$G = \langle\, a, b \mid a^7 = (a^2 b)^3 = (a^3 b)^2 = (ab^5)^2 = 1 \,\rangle.$$

Prove that $G$ may also be presented as

$$\langle\, x, y \mid x^2 = y^3 = (xy)^7 = ((y^{-1} x y x)^4 y^{-1} x)^2 = 1 \,\rangle.$$

**4.15**   Let $G$ be the group with presentation

$$\langle\, a, b, c, d, e \mid ab = c,\ bc = d,\ cd = e,\ de = a,\ ea = b \,\rangle.$$

By eliminating $c, d, e$ show that

(1) $\qquad\qquad\qquad\qquad a = babab^2 ab,$

(2) $\qquad\qquad\qquad\qquad b = ab^2 aba.$

Replacing the final $a$ in (2) by the expression given by (1), show that $b^5 = a^{-2}$. Deduce, by multiplying (1) on the right by $a$ and using (2), that $a = b^{-8}$.

Conclude that $G \simeq C_{11}$.

**4.16**   Show that the group

$$G = \langle\, a, b \mid ab = b^2 a,\ ba = a^2 b \,\rangle$$

is the trivial group.

     More generally, consider the group

$$G_n = \langle\, a, b \mid ab^n = b^{n+1} a,\ ba^n = a^{n+1} b \,\rangle.$$

Prove by induction on $i$ that

$$a^i b^{n^i} a^{-i} = b^{(n+1)^i}.$$

Using the relations obtained by taking $i = n$ and $i = n + 1$, deduce that $G_n$ is the trivial group.

**4.17**   Let $\mathrm{SL}(2, 7)$ denote the group of $2 \times 2$ matrices of determinant 1 with entries in $\mathbb{Z}_7$, and let

$$\mathrm{PSL}(2, 7) = \mathrm{SL}(2, 7)/Z(\mathrm{SL}(2, 7)).$$

Let $\natural : \mathrm{SL}(2, 7) \to \mathrm{PSL}(2, 7)$ be the natural map and let

$$a = \natural \begin{bmatrix} 0 & 6 \\ 1 & 0 \end{bmatrix}, \quad b = \natural \begin{bmatrix} 5 & 6 \\ 6 & 6 \end{bmatrix}.$$

Show that $\langle\, a, b \,\rangle$ is the dihedral group of order 8.

# 4: Presentations

**4.18** Let GL(2,3) be the group of $2 \times 2$ non-singular matrices with entries in $\mathbb{Z}_3$. Show that $|\text{GL}(2,3)| = 48$.

Prove that SL(2,3) is the derived group of GL(2,3).

The quaternion group $Q_8$ may be presented by

$$Q_8 = \langle\, a, b, c \mid ab = c,\ bc = a,\ ca = b \,\rangle.$$

From this presentation it is clear that $Q_8$ has an automorphism $\vartheta$ of order 3 which permutes $a, b, c$ cyclically. Let $H$ be $Q_8$ extended by this automorphism $\vartheta$ of order 3. Show that $H$ is isomorphic to SL(2,3). Hence show that GL(2,3) has derived length 4.

**4.19** Let $H$ be the group with presentation

$$\langle\, a, b \mid a^2 = b^3 = 1,\ (ab)^n = (ab^{-1}ab)^k \,\rangle.$$

Show that $H$ is generated by $ab$ and $ab^{-1}ab$. Deduce that $\langle (ab)^n \rangle$ is contained in the centre of $H$. Prove also that $\langle (ab)^n \rangle$ is contained in the derived group of $H$.

**4.20** Let $G$ be the group with two generators $a, b$ subject to the relations $x^3 = 1$ for all $x \in G$. Show that $[a, b]$ belongs to the centre of $G$. Deduce that $G$ is finite.

**4.21** For any integers $a, b, c$ define a group $G$ by

$$G = \langle\, x, y \mid x^2 = 1,\ xy^a xy^b xy^c = 1 \,\rangle.$$

Prove that $y^a(xy^{b-c}x)y^{-b} = xy^{a-c}x$ and find two similar relations with $a, b, c$ permuted cyclically.

Deduce that $y^{2(a+b+c)}$ commutes with $xy^{a-c}x$. Prove also that if h.c.f.$(a - c, b - c) = 1$ then $y^{2(a+b+c)}$ commutes with $xyx$.

Finally, show that if h.c.f.$(a - c, b - c) = 1$ then

$$y^{2(a+b+c)} \in Z(G).$$

**4.22** Let $G$ be the group

$$\langle\, x, t \mid xt^{m+1} = t^2 x^2,\ xt^2 xtx^2 t = 1 \,\rangle.$$

Prove that

(a) $xt^{2m+2}x^{-1} = tx^{-2}t^{-1}$;

(b) $[t^2, xtx] = 1$;

(c) $xt^2 x^{-1} = t^{2m+1}x^{-2}t^{-1}$;

(d) $xt^{2m}x^{-1} = t^{-2m}$.

Deduce that $t^{2m} \in Z(G)$ and $t^{4m} = 1$.

27

**4.23** Let $SL(2, \mathbf{Z})$ be the group of $2 \times 2$ matrices of determinant 1 with entries in $\mathbf{Z}$. Let

$$ s = \begin{bmatrix} 1 & 1 \\ 0 & 1 \end{bmatrix}, \quad t = \begin{bmatrix} 0 & -1 \\ 1 & 0 \end{bmatrix}. $$

Prove that if $\begin{bmatrix} a & b \\ c & 0 \end{bmatrix} \in SL(2, \mathbf{Z})$ then $\begin{bmatrix} a & b \\ c & 0 \end{bmatrix} \in \langle s, t \rangle$.

Suppose, by way of an inductive hypothesis, that $\begin{bmatrix} a' & b' \\ c' & d' \end{bmatrix} \in SL(2, \mathbf{Z})$ with $|d'| < |d|$ implies $\begin{bmatrix} a' & b' \\ c' & d' \end{bmatrix} \in \langle s, t \rangle$. If

$$ m = \begin{bmatrix} a & b \\ c & d \end{bmatrix} \in SL(2, \mathbf{Z}) $$

prove, by considering $t s^n m$ where $|b + nd| < |d|$, that $m \in \langle s, t \rangle$. Deduce that $SL(2, \mathbf{Z}) = \langle s, t \rangle$.

Now let $u = st$ and denote by $\bar{u}, \bar{t}$ the images of $u, t$ under the natural map $\natural : SL(2, \mathbf{Z}) \to PSL(2, \mathbf{Z})$. Use the above results to show that $PSL(2, \mathbf{Z}) = \langle \bar{u}, \bar{t} \rangle$. Show also that $\bar{u}^3 = \bar{t}^2 = \bar{I}$, the identity of $PSL(2, \mathbf{Z})$.

Suppose, if possible, that some word of the form

$$ \cdots \bar{u}^{\pm 1} \bar{t} \bar{u}^{\pm 1} \bar{t} \bar{u}^{\pm 1} \bar{t} \cdots $$

is equal to $\bar{I}$ in $PSL(2, \mathbf{Z})$. Show that there is a word of the form

$$ w = u^{\pm 1} t u^{\pm 1} \cdots t u^{\pm 1} t $$

that is equal to $\pm I$ in $SL(2, \mathbf{Z})$ and, by considering the trace of $w$, obtain a contradiction.

Finally, show that

$$ PSL(2, \mathbf{Z}) \simeq \langle a, b \mid a^2 = b^3 = 1 \rangle. $$

# Solutions to Chapter 1

**1.1** Let $G$ be partitioned by the set $\{Hx_\alpha \mid \alpha \in A\}$ of cosets of $H$, and let $H$ be partitioned by the set $\{Ky_\beta \mid \beta \in B\}$ of cosets of $K$. Suppose that $g \in G$. Then we have $g \in Hx_\alpha$ for some $\alpha \in A$ and so $g = hx_\alpha$ for some (unique) $h \in H$. But $h \in Ky_\beta$ for some $\beta \in B$ and so we have that $g = ky_\beta x_\alpha$ for some $k \in K$. Thus we see that every element of $G$ belongs to a coset $Ky_\beta x_\alpha$ for some $\beta \in B$ and some $\alpha \in A$. The result now follows from the fact that if $Ky_\beta x_\alpha = Ky_{\beta'} x_{\alpha'}$ then, since the left hand side is contained in the coset $Hx_\alpha$ and the right hand side is contained in the coset $Hx_{\alpha'}$, we have necessarily $x_\alpha = x_{\alpha'}$, which gives $Ky_\beta = Ky_{\beta'}$ and hence $y_\beta = y_{\beta'}$.

Now observe that $(H \cap K)x = Hx \cap Kx$ for all subgroups $H$ and $K$ of $G$. Then, if $H$ and $K$ have finite index, the fact that there are finitely many cosets $Hx$ and $Kx$ implies that there are only finitely many cosets of $H \cap K$, so $H \cap K$ is also of finite index. The result for the intersection of a finite number of subgroups now follows by induction.

The result is not true for an infinite number of subgroups each of finite index. To see this, consider the additive group $\mathbb{Z}$. The subgroup $n\mathbb{Z}$ has index $n$; but $\bigcap_{n \geq 1} n\mathbb{Z} = \{0\}$ which is not of finite index in $\mathbb{Z}$.

**1.2** If $xH$ is a subgroup of $G$ then we have $1 \in xH$ which gives $x^{-1} \in x^{-1}xH = H$ and hence $x \in H$, so that $xH = H$.

That $\varphi$ is a mapping (or, as some say, is well-defined) follows from the observation that

$$xH = yH \implies y^{-1}x \in H$$
$$\implies Hy^{-1}x = H$$
$$\implies Hy^{-1} = Hx^{-1}.$$

It is clear that $\varphi$ is a bijection.

$\psi$ on the other hand is not a mapping. To see this, take for example $G = \mathrm{GL}(2, \mathbb{Q})$ and

$$H = \left\{ \begin{bmatrix} a & b \\ 0 & c \end{bmatrix} \mid a, b, c \in \mathbb{Q}, \ ac \neq 0 \right\}.$$

We have that

$$\begin{bmatrix} 1 & 0 \\ 1 & 1 \end{bmatrix} H = \begin{bmatrix} 1 & 1 \\ 1 & 2 \end{bmatrix} H,$$

which is immediate from the observation that if

$$A = \begin{bmatrix} 1 & 0 \\ 1 & 1 \end{bmatrix}, \quad B = \begin{bmatrix} 1 & 1 \\ 1 & 2 \end{bmatrix}$$

then

$$A^{-1}B = \begin{bmatrix} 1 & 1 \\ 0 & 1 \end{bmatrix} \in H.$$

However,

$$H \begin{bmatrix} 1 & 0 \\ 1 & 1 \end{bmatrix} \neq H \begin{bmatrix} 1 & 1 \\ 1 & 2 \end{bmatrix},$$

since equality here would give the contradiction

$$BA^{-1} = \begin{bmatrix} 0 & 1 \\ -1 & 2 \end{bmatrix} \in H.$$

**1.3**    Consider the subgroups $H$ and $K$ of the group $G = \mathrm{GL}(2, \mathbb{Q})$ given by

$$H = \left\{ \begin{bmatrix} 1 & b \\ 0 & 1 \end{bmatrix} \mid b \in \mathbb{Q} \right\}, \quad K = \left\{ \begin{bmatrix} 1 & 0 \\ a & 1 \end{bmatrix} \mid a \in \mathbb{Q} \right\}.$$

It is readily seen that

$$HK = \left\{ \begin{bmatrix} 1 + ab & b \\ a & 1 \end{bmatrix} \mid a, b \in \mathbb{Q} \right\}.$$

But $HK$ is not a subgroup of $G$ since, for example, the matrices

$$\begin{bmatrix} 2 & 1 \\ 1 & 1 \end{bmatrix}, \quad \begin{bmatrix} 1 & 1 \\ 0 & 1 \end{bmatrix}$$

are each in $HK$ but the product

$$\begin{bmatrix} 2 & 1 \\ 1 & 1 \end{bmatrix} \begin{bmatrix} 1 & 1 \\ 0 & 1 \end{bmatrix} = \begin{bmatrix} 2 & 3 \\ 1 & 2 \end{bmatrix}$$

is not.

## Solutions to Chapter 1

**1.4** The right cosets are

$$H(1) = \{(1), (12)\}$$
$$H(13) = \{(13), (132)\}$$
$$H(23) = \{(23), (123)\}.$$

The left cosets are

$$(1)H = \{(1), (12)\}$$
$$(13)H = \{(13), (123)\}$$
$$(23)H = \{(23), (132)\}.$$

It is clear from this that $H$ is not normal in $G$.

**1.5** Since $m$ and $n$ are coprime there exist integers $a$ and $b$ with $am+bn = 1$. Then, using the fact that $g^n = 1$, we have

$$g = g^1 = g^{am+bn} = g^{am}g^{bn} = (g^m)^a(g^n)^b = (g^m)^a.$$

Since it is given that $g^m \in H$ we have that $(g^m)^a \in H$ and hence $g \in H$.

**1.6** (i) If $H$ is a subgroup of $G$ then clearly $HH \subseteq H$; and, since every subgroup contains the identity element, we have $H = 1H \subseteq HH$.

(ii) Given $x \in X$ we have $xX \subseteq XX = X$. Since $y \mapsto xy$ is injective (by the cancellation law) we deduce that $|xX| = |X|$ and hence, since $X$ is finite, that $xX = X$. Consequently, $x = xe$ for some $e \in X$. The cancellation law gives $e = 1$, and so we have that $1 \in X$. We now observe from $1 \in xX$ that $1 = xy$ for some $y \in X$, which gives $x^{-1} = y \in X$. It now follows from the fact that $XX \subseteq X$ that $X$ is a subgroup of $G$.

That (ii) no longer holds when $X$ is infinite may be seen, for example, by taking $G$ to be the additive group of integers and $X$ to be the set of non-negative integers.

**1.7** Suppose that $t \in HxK \cap yK \neq \emptyset$. Then (with a notation that is self-explanatory) we have $t = hxk$ and $t = yk_1$. If now $s \in yK$ then

$$s = yk_2 = tk_1^{-1}k_2 = hxkk_1^{-1}k_2 \in HxK,$$

and hence $yK \subseteq HxK$.

Suppose now that $t \in HxK \cap HyK \neq \emptyset$. Then we have $t = h_1xk_1$ and $t = h_2yk_2$. If now $s \in HxK$ then

$$s = hxk = h(h_1^{-1}tk_1^{-1})k = hh_1^{-1}(h_2yk_2)k_1^{-1}k \in HyK.$$

Thus $HxK \subseteq HyK$, and similarly $HyK \subseteq HxK$.

1.8    Let $C_n = \langle a \rangle$ with $n = p^m$ where $p$ is prime. If $H$ and $K$ are subgroups
of $C_n$ then we have $H = \langle a^s \rangle$ and $K = \langle a^t \rangle$. Now we can assume that
$s = p^u$ and $t = p^v$ where $0 \le u \le m$ and $0 \le v \le m$. For, if $s = kp^u$
where h.c.f.$(k,p) = 1$ then there are integers $x, y$ with $xk + yp^m = 1$.
Then $(a^{kp^u})^x \in H$ and $(a^{p^m})^{yp^u} \in H$ since $a^{p^m} = 1$. Thus

$$a^{xkp^u + p^{m+u}y} = a^{p^u} \in H$$

and so $H = \langle a^{p^u} \rangle$. It now follows that if $u \le v$ then we have $K \subseteq H$,
while if $v \le u$ then $H \subseteq K$.

The converse is also true. Suppose that $n = pq$ where h.c.f.$(p,q) = 1$.
Then if $C_n = \langle a \rangle$ we have that $|\langle a^p \rangle| = q$ and $|\langle a^q \rangle| = p$. Since $p$ and $q$
are coprime, we have $\langle a^p \rangle \cap \langle a^q \rangle = \{1\}$. But this is not possible under
the assumption that, for any two subgroups $H$ and $K$, either $H \subseteq K$ or
$K \subseteq H$.

1.9    Since $g^{-1}Hg$ is a subgroup of $G$ for every $g \in G$ we have that $H_G = \bigcap_{g \in G} g^{-1}Hg$ is a subgroup of $G$. Let $x \in H_G$. Then $x \in g^{-1}Hg$ for
every $g \in G$ and so, for every $y \in G$, we have

$$y^{-1}xy \in y^{-1}g^{-1}Hgy = (gy)^{-1}Hgy$$

which shows that $y^{-1}xy \in H_G$. Thus we see that $H_G$ is a normal
subgroup of $G$.

Suppose now that $K$ is a subgroup of $H$ that is normal in $G$. Then if
$k \in K$ we have $gkg^{-1} \in K$ for every $g \in G$ and so

$$k \in g^{-1}Kg \subseteq g^{-1}Hg$$

which shows that $k \in \bigcap_{g \in G} g^{-1}Hg = H_G$.

In the case where $G = \mathrm{GL}(2, \mathbb{Q})$ and $H$ is the subgroup of non-singular
diagonal matrices, i.e.

$$H = \left\{ \begin{bmatrix} a & 0 \\ 0 & b \end{bmatrix} \mid a, b \in \mathbb{Q},\ ab \ne 0 \right\},$$

consider the subset $K$ of $H$ described by

$$K = \left\{ \begin{bmatrix} a & 0 \\ 0 & a \end{bmatrix} \mid a \in \mathbb{Q} \backslash \{0\} \right\}.$$

Clearly, $K$ is a normal subgroup of $G$ and so, by the above, we have $K \subseteq H_G$. But if $x = \begin{bmatrix} a & 0 \\ 0 & b \end{bmatrix} \in H_G$ then from

$$\begin{bmatrix} 1 & -1 \\ 0 & 1 \end{bmatrix}\begin{bmatrix} a & 0 \\ 0 & b \end{bmatrix}\begin{bmatrix} 1 & 1 \\ 0 & 1 \end{bmatrix} = \begin{bmatrix} a & a-b \\ 0 & b \end{bmatrix} \in H_G \subseteq H$$

we deduce that $a - b = 0$ whence $x \in K$ and so $H_G = K$.

$K$ is isomorphic to the group of non-zero rationals under multiplication.

**1.10** That $H$ is a group is routine. The only element of order 2 is $\begin{bmatrix} -1 & 0 \\ 0 & -1 \end{bmatrix}$.

$$\left\langle \begin{bmatrix} 0 & 1 \\ -1 & 0 \end{bmatrix} \right\rangle = \left\langle \begin{bmatrix} 0 & -1 \\ 1 & 0 \end{bmatrix} \right\rangle \simeq C_4$$

$$\left\langle \begin{bmatrix} 0 & i \\ i & 0 \end{bmatrix} \right\rangle = \left\langle \begin{bmatrix} 0 & -i \\ -i & 0 \end{bmatrix} \right\rangle \simeq C_4$$

$$\left\langle \begin{bmatrix} -i & 0 \\ 0 & i \end{bmatrix} \right\rangle = \left\langle \begin{bmatrix} i & 0 \\ 0 & -i \end{bmatrix} \right\rangle \simeq C_4$$

$\left\langle \begin{bmatrix} -1 & 0 \\ 0 & -1 \end{bmatrix} \right\rangle \simeq C_2$ and is a subgroup of all the three cyclic subgroups of order 4 given above.

The only other subgroups are $H$ and $\{1\}$. For, if $K$ is a subgroup of $H$ then $|K|$ divides $|H| = 8$, so $|K|$ is 1, 2, 4, or 8. The subgroups of orders 1,8 are $\{1\}, H$ respectively. There is only one subgroup of order 2 since there is only one element of order 2. If $|K| = 4$ then either $K$ is cyclic (and so is given above) or $K$ has all its non-trivial elements of order 2 (since the order of an element divides the order of the group). This is impossible since there is only one element of order 2. All these subgroups are normal (although the group is not abelian). $H$ cannot have a quotient group isomorphic to $C_4$. For, if $H/K \simeq C_4$ then $K \simeq C_2$ so $K = \left\langle \begin{bmatrix} -1 & 0 \\ 0 & -1 \end{bmatrix} \right\rangle$. But every non-trivial element has square equal to $\begin{bmatrix} -1 & 0 \\ 0 & -1 \end{bmatrix}$ so every non-trivial coset $xK$ has order 2. Consequently $H/K \not\simeq C_4$.

**1.11**   Let $a = \begin{bmatrix} 0 & 1 \\ 1 & 0 \end{bmatrix}$ and $b = \begin{bmatrix} \alpha & 0 \\ 0 & \alpha^{-1} \end{bmatrix}$ where $\alpha = e^{2\pi i/n}$. Then clearly $a$ is of order 2 and $b$ is of order $n$. Moreover,

$$a^{-1}ba = \begin{bmatrix} 0 & 1 \\ 1 & 0 \end{bmatrix} \begin{bmatrix} \alpha & 0 \\ 0 & \alpha^{-1} \end{bmatrix} \begin{bmatrix} 0 & 1 \\ 1 & 0 \end{bmatrix}$$

$$= \begin{bmatrix} \alpha^{-1} & 0 \\ 0 & \alpha \end{bmatrix}$$

$$= b^{-1}.$$

Thus $\langle b \rangle$ is a normal subgroup of $D_{2n}$. Now

$$D_{2n}/\langle b \rangle = \{\langle b \rangle, \; a\langle b \rangle\}$$

and so $|D_{2n}/\langle b \rangle| = 2$ and $|D_{2n}| = 2n$. The subgroup $\langle b \rangle$ is cyclic and of index 2.

We observe that $|G| = 2n$; for $\epsilon$ takes two possible values and $k$ takes $n$ possible values. Now it is readily seen that the assignment

$$a \mapsto \begin{bmatrix} -1 & 0 \\ 0 & 1 \end{bmatrix}, \qquad b \mapsto \begin{bmatrix} 1 & 1 \\ 0 & 1 \end{bmatrix}$$

sets up an isomorphism between $D_{2n}$ and the subgroup of $G$ that is generated by

$$\left\{ \begin{bmatrix} -1 & 0 \\ 0 & 1 \end{bmatrix}, \begin{bmatrix} 1 & 1 \\ 0 & 1 \end{bmatrix} \right\}.$$

Since this subgroup has order $2n$, which is the order of $G$, it must coincide with $G$ and so $G$ is isomorphic to $D_{2n}$.

Consider now the mapping from $D_\infty \to D_{2n}$ described by the assignment

$$\begin{bmatrix} \epsilon & k \\ 0 & 1 \end{bmatrix} \mapsto \begin{bmatrix} \epsilon & k \;(\mathrm{mod}\; n) \\ 0 & 1 \end{bmatrix}.$$

It is readily seen that this is a group morphism. Since it is clearly surjective, it follows by the first isomorphism theorem that $D_{2n}$ is a quotient group of $D_\infty$.

## Solutions to Chapter 1

**1.12**    (i) Define $f : \mathbb{C}^+ \to \mathbb{R}^+$ by $f(a+ib) = a$. Then $f$ is a group morphism which is surjective. Since

$$\mathrm{Ker}\, f = \{a + ib \in \mathbb{C}^+ \mid a = 0\} \simeq \mathbb{R}^+,$$

the result follows by the first isomorphism theorem.

(ii) Define $f : \mathbb{C}^\bullet \to U$ by

$$a + ib \mapsto \frac{a}{\sqrt{a^2 + b^2}} + i\frac{b}{\sqrt{a^2 + b^2}}.$$

Then $f$ is a surjective group morphism with

$$\mathrm{Ker}\, f = \{a + ib \in \mathbb{C}^\bullet \mid b = 0,\ \frac{a}{\sqrt{a^2}} = 1\} \simeq \mathbb{R}^\bullet_{>0}.$$

The result now follows by the first isomorphism theorem.

(iii) Define $f : \mathbb{C}^\bullet \to \mathbb{R}^\bullet_{>0}$ by $a + ib \mapsto \sqrt{a^2 + b^2}$, and define $g : \mathbb{R}^\bullet \to \mathbb{R}^\bullet_{>0}$ by $a \mapsto |a|$. Then $f$ and $g$ are surjective group morphisms and the result follows from the observation that

$$\mathrm{Ker}\, f = \{a + ib \in \mathbb{C}^\bullet \mid \sqrt{a^2 + b^2} = 1\} \simeq U;$$
$$\mathrm{Ker}\, g = \{1, -1\} \simeq C_2.$$

(iv) Define $f : \mathbb{R}^\bullet \to C_2$ by

$$f(a) = \begin{cases} 1 & \text{if } a > 0; \\ -1 & \text{if } a < 0. \end{cases}$$

Then $f$ is a surjective group morphism with $\mathrm{Ker}\, f = \mathbb{R}^\bullet_{>0}$.

Also, define $g : \mathbb{Q}^\bullet \to C_2$ by

$$g(a) = \begin{cases} 1 & \text{if } a > 0; \\ -1 & \text{if } a < 0. \end{cases}$$

Then $g$ is a surjective group morphism with $\mathrm{Ker}\, g = \mathbb{Q}^\bullet_{>0}$.

(v) Define $f : \mathbb{Q}^\bullet \to \mathbb{Q}^\bullet_{>0}$ by $a \mapsto |a|$.

**1.13**    We have the chain of subgroups

$$\{1\} \subseteq C_p \subseteq C_{p^2} \subseteq C_{p^3} \subseteq \cdots \subseteq C_{p^n} \subseteq \cdots \subseteq \mathbb{Z}_{p^\infty}$$

with $\mathbb{Z}_{p^\infty} = \bigcup_{n \geq 1} C_{p^n}$. Here the cyclic subgroup $C_{p^n}$ is generated by a primitive $p^n$th root of unity. To see that every proper subgroup of

35

$\mathbf{Z}_{p^\infty}$ is cyclic note that, given any subset $X$ of $\mathbf{Z}_{p^\infty}$, there is a smallest member of the chain that contains $X$ (this, of course, might be $\mathbf{Z}_{p^\infty}$ itself). Now if $X$ is finite then it is clear that this smallest member is $C_{p^n}$ for some $n$; but if $X$ is infinite then it must contain $p^n$th roots of unity for arbitrarily large $n$ and so must generate $\mathbf{Z}_{p^\infty}$.

To see that $\mathbf{Z}_{p^\infty}/C_{p^n} \simeq \mathbf{Z}_{p^\infty}$, consider the mapping $f : \mathbf{Z}_{p^\infty} \to \mathbf{Z}_{p^\infty}$ described by

$$z \mapsto z^{p^n}.$$

It is readily seen that $f$ is a surjective group morphism with $\mathrm{Ker}\, f = C_{p^n}$, so the result follows by the first isomorphism theorem.

For $\mathbf{Q}^+$, suppose that

$$X = \left\{ \frac{p_1}{q_1}, \frac{p_2}{q_2}, \cdots , \frac{p_n}{q_n} \right\}.$$

Then clearly we have that

$$X \subseteq \left\langle \frac{1}{q_1 q_2 \cdots q_n} \right\rangle$$

which is a cyclic group. Hence $\langle X \rangle$ is a subgroup of a cyclic group and therefore is itself a cyclic group.

1.14  If no element of $G$ has order 4 then clearly every element has order 1 or 2 and so $x^2 = 1$ for every $x \in G$. But then $(xy)^2 = 1$ gives

$$xy = (xy)^{-1} = y^{-1}x^{-1} = y^2 y^{-1} x^{-1} x^2 = yx$$

and so the group is abelian.

It follows from this that if $G$ is non-abelian of order 8 then $G$ contains an element of order 4. Let this element be $a$. Clearly, $\{1, a, a^2, a^3\}$ has index 2 in $G$ and so is normal. Suppose that $b \neq a^2$ is of order 2 in $G$. Then $b^{-1}ab \neq a$ (otherwise $G$ would be abelian) and so $b^{-1}ab$ is of order 4 in $\langle a \rangle$ and so must be $a^3$. Thus $G$ is the group

$$\langle a, b \mid a^4 = b^2 = 1,\ b^{-1}ab = a^{-1} \rangle$$

which has order 8.

If $G$ does not contain any other element of order 2 we can choose $b \notin \langle a \rangle$ of order 4. Then $b^2$ has order 2 and so $b^2 = a^2$. As in the above, $b^{-1}ab \neq a$ and so $b^{-1}ab = a^3$ and $G$ is the quaternion group

$$\langle a, b \mid a^4 = b^4 = 1,\ a^2 = b^2,\ b^{-1}ab = a^{-1} \rangle.$$

These groups are not isomorphic since the number of elements of order 2 in each is different.

# Solutions to Chapter 1

**1.15** The possible orders are the positive divisors of 24, namely 1, 2, 3, 4, 6, 8, 12, and 24.

| | | |
|---|---|---|
| (1) | 1 element | order 1 |
| (12) | 6 elements | order 2 |
| (123) | 8 elements | order 3 |
| (1234) | 6 elements | order 4 |
| (12)(34) | 3 elements | order 2 |

There is one element of order 1; nine elements of order 2; eight elements of order 3; six elements of order 4. There are nine subgroups of order 2, and four subgroups of order 3 (each of which contains two elements of order 3).

The cyclic subgroups of order 4 are

$$\{(1), (1234), (13)(24), (1432)\}$$
$$\{(1), (1324), (12)(34), (1423)\}$$
$$\{(1), (1243), (14)(23), (1342)\}.$$

The non-cyclic subgroups of order 4 ($\simeq C_2 \times C_2$) are

$$\{(1), (12)(34), (13)(24), (14)(23)\}$$
$$\{(1), (13), (24), (13)(24)\}$$
$$\{(1), (14), (23), (14)(23)\}.$$

The subgroups of order 6 are those that fix one of 1, 2, 3, or 4. Hence there are four such subgroups; for example, 4 is fixed by

$$\{(1), (12), (13), (23), (123), (132)\}.$$

There are three subgroups of order 8, namely

$$\{(1), (13), (24), (13)(24), (12)(34), (14)(23), (1234), (1432)\}$$
$$\{(1), (12), (34), (12)(34), (13)(24), (14)(23), (1324), (1423)\}$$
$$\{(1), (14), (23), (14)(23), (12)(34), (13)(24), (1243), (1342)\}.$$

$A_4$ is a subgroup of $S_4$ of order 12.
An abelian normal subgroup of $S_4$ is

$$V = \{(1), (12)(34), (13)(24), (14)(23)\}.$$

$S_4/V$ has order 6 and is isomorphic to $S_3$ (which is a subgroup of $S_4$).

A subgroup of $A_4$ is necessarily a subgroup of $S_4$. Hence, if $A_4$ had a subgroup $H$ with $|H| = 6$ then $H$ must be one of the subgroups of $S_4$ of order 6. But none of these consist only of even permutations. Hence $A_4$ has no subgroup of order 6. This, incidentally, shows that the converse of Lagrange's theorem is false.

*1.16*   We have that

$$
\begin{aligned}
a &= (1234)(5678) \\
a^2 &= (13)(24)(57)(68) \\
a^3 &= (1432)(5876) \\
a^4 &= (1) \\
b &= (1537)(2846) \\
b^2 &= (13)(57)(24)(86) = a^2 \\
b^3 &= (1735)(2648) \\
b^4 &= (1) = a^4 \\
ab &= (1836)(2745) \\
ba &= (1638)(2547) \\
(ab)^2 &= (13)(86)(24)(75) = a^2 \\
(ab)^3 &= (1638)(2547) = ba.
\end{aligned}
$$

The assignment

$$
\begin{bmatrix} 0 & 1 \\ -1 & 0 \end{bmatrix} \longmapsto (1234)(5678)
$$

$$
\begin{bmatrix} 0 & i \\ i & 0 \end{bmatrix} \longmapsto (1537)(2846)
$$

extends to give the following isomorphism :

$$
\begin{bmatrix} 1 & 0 \\ 0 & 1 \end{bmatrix} \longmapsto (1)
$$

$$
\begin{bmatrix} -1 & 0 \\ 0 & -1 \end{bmatrix} \longmapsto (13)(24)(57)(68)
$$

$$
\begin{bmatrix} 0 & 1 \\ -1 & 0 \end{bmatrix} \longmapsto (1234)(5678)
$$

$$
\begin{bmatrix} 0 & -1 \\ 1 & 0 \end{bmatrix} \longmapsto (1432)(5876)
$$

$$
\begin{bmatrix} 0 & i \\ i & 0 \end{bmatrix} \longmapsto (1537)(2846)
$$

## Solutions to Chapter 1

$$\begin{bmatrix} 0 & -1 \\ -1 & 0 \end{bmatrix} \longmapsto (1735)(2648)$$

$$\begin{bmatrix} -i & 0 \\ 0 & i \end{bmatrix} \longmapsto (1638)(2547)$$

$$\begin{bmatrix} i & 0 \\ 0 & -i \end{bmatrix} \longmapsto (1836)(2745).$$

It is easy to see that the quaternion group is not isomorphic to any of the subgroups of order 8 in $S_4$. For, in the quaternion group, there is only one element of order 2 whereas the subgroups of order 8 in $S_4$ have five elements of order 2. The subgroups of order 8 in $S_4$ are all isomorphic to the dihedral group $D_8$. These are, in fact, the only two non-abelian groups of order 8 (see question 1.14).

**1.17**  Suppose that

$$p = (a_{11} \cdots a_{1m})(a_{21} \cdots a_{2m}) \cdots (a_{r1} \cdots a_{rm}).$$

Then we have that $p = \vartheta^r$ where

$$\vartheta = (a_{11}\, a_{21} \cdots a_{r1}\, a_{12}\, a_{22} \cdots a_{r2} \cdots).$$

Conversely, suppose that $\vartheta = (1\, 2 \cdots m)$. Then

$$\vartheta^s = (1 \quad s+1 \quad 2s+1 \quad \cdots \quad (k-1)s+1)$$
$$\circ (2 \quad s+2 \quad 2s+2 \quad \cdots \quad (k-1)s+2)$$
$$\circ \cdots$$
$$\circ (\frac{m}{k} \quad s+\frac{m}{k} \quad 2s+\frac{m}{k} \quad \cdots \quad (k-1)s+\frac{m}{k})$$

where $k$ is the least positive integer such that $ks$ is divisible by $m$. Hence $k = m/\text{h.c.f.}(m,s)$ and the cycles have length $m/\text{h.c.f.}(m,s)$. Also, there are $m/k = \text{h.c.f.}(m,s)$ cycles in the decomposition as required.

**1.18**  We can choose $a$ in $p-1$ ways since $\mathbb{Z}_p$ contains $p-1$ non-zero elements. The elements $b$ and $c$ are arbitrary while $d$ is uniquely determined by the condition that $ad - bc = 1$ (i.e. $d = a^{-1}(1 + bc)$). Thus there are $p^2(p-1)$ elements of this form.

Consider now $\begin{bmatrix} 0 & b \\ c & d \end{bmatrix}$. As $d$ is arbitrary it can be chosen in $p$ ways. Since $-bc = 1$ the element $b$ must be non-zero, so can be chosen in

39

$p - 1$ ways. Then $c = -b^{-1}$ is uniquely determined. There are therefore $p(p - 1)$ elements of this form.

We thus have

$$|SL(2, p)| = p^2(p - 1) + p(p - 1) = p(p - 1)(p + 1).$$

The centre $Z$ of $SL(2, p)$ is

$$\left\{ \begin{bmatrix} 1 & 0 \\ 0 & 1 \end{bmatrix}, \begin{bmatrix} -1 & 0 \\ 0 & -1 \end{bmatrix} \right\}.$$

Hence, if $p \neq 2$, we have $|Z| = 2$ and so

$$|SL(2, p)/Z| = \tfrac{1}{2}p(p - 1)(p + 1).$$

Since $SL(2, 2)$ is a group of order 6 and is non-abelian, it must be the symmetric group $S_3$.

We count the number of $n \times n$ matrices over $\mathbb{Z}_p$ with linearly independent rows. The first row can be any $n$–tuple except zero, so there are $p^n - 1$ possible first rows. Now there are $p$ multiples of the first row and the second row can be any except these; so there are $p^n - p$ possible second rows. Again, there are $p^2$ linear combinations of the first two rows and the third row can be any but these; so there are $p^n - p^2$ possible third rows. Continuing in this way, we see that there are

$$\prod_{i=0}^{n-1}(p^n - p^i)$$

such matrices.

Finally, consider the morphism from the group of these matrices to the multiplicative group of non-zero elements of $\mathbb{Z}_p$ that is described by the determinant map. The result follows from the fact that $SL(n, p)$ is the kernel of this morphism.

**1.19** Suppose that $A = \begin{bmatrix} a & b \\ c & d \end{bmatrix}$ and that $\operatorname{tr}(A) = 0$, i.e. $d = -a$. Then since

$$A = \begin{bmatrix} a & b \\ c & -a \end{bmatrix} \in SL(2, F)$$

40

# Solutions to Chapter 1

we have that $-a^2 - bc = 1$. Consequently,

$$A^2 = \begin{bmatrix} a^2 + bc & 0 \\ 0 & bc + a^2 \end{bmatrix} = -I_2.$$

Conversely, suppose that $A \in \mathrm{SL}(2, F)$ is such that $A^2 = -I_2$. Then if $A = \begin{bmatrix} a & b \\ c & d \end{bmatrix}$ we have $ad - bc = 1$ and

$$\begin{bmatrix} a^2 + bc & ab + bd \\ ca + dc & cb + d^2 \end{bmatrix} = \begin{bmatrix} -1 & 0 \\ 0 & -1 \end{bmatrix}.$$

Suppose that $a + d \neq 0$. Then since $(a+d)b = 0$ and $(a+d)c = 0$ we have $b = c = 0$. From $ad - bc = 1$ we then have $ad = 1$. But $a + d \neq 0$ so $a^2 + ad = a(a+d) \neq 0$. This contradicts $a^2 = -1$ and $ad = 1$. Thus $a + d = 0$ as required.

Now if $\mathrm{tr}\,(A) = 0$ then $A^2 = -I_2$ and hence $\overline{A}^2$ is the identity since

$$Z(\mathrm{SL}(2, F)) = \left\{ \begin{bmatrix} 1 & 0 \\ 0 & 1 \end{bmatrix}, \begin{bmatrix} -1 & 0 \\ 0 & -1 \end{bmatrix} \right\}.$$

Conversely, if $\overline{A}^2$ is the identity of $\mathrm{PSL}(2, F)$ then we have either $A^2 = I_2$ or $A^2 = -I_2$. If $A^2 = -I_2$ then $\mathrm{tr}\,(A) = 0$ as required. If $A^2 = I_2$ then it is easily seen that either $A = I_2$ or $A = -I_2$, and in either case $\overline{A}$ is the identity of $\mathrm{PSL}(2, F)$ so is not an element of order 2.

1.20  Every element of $C_2 \times C_2$ has order 2 so, since $|C_2 \times C_2| = 4$, it follows that $C_2 \times C_2$ cannot be cyclic.

Suppose now that $G$ is non-cyclic and of order 4. Then every element of $G$ has order 2 and the multiplication table is uniquely determined. Since this is the same as that of $C_2 \times C_2$ the result follows.

1.21  If $p \neq q$ then $C_p \times C_q \simeq C_{pq}$ and, since a cyclic group has only one subgroup of each order and the order of a subgroup divides the order of the group, there can be only two subgroups of $C_{pq}$ other than $\{1\}$ and $C_{pq}$.

However, $C_p \times C_p$ has more than two proper non-trivial subgroups; for $\{1\} \times C_p, C_p \times \{1\}$ and $\langle (a, b) \rangle$ (where $a, b$ are non-trivial) are all isomorphic to $C_p$.

**1.22** Let $K$ be a normal subgroup of $G \times H$ with $K \neq \{(1,1)\}$. Suppose that $(x,1) \in K$ for some $x \neq 1$ in $G$. Then for every $g \in G$ we have, since $K$ is normal,

$$(g^{-1}xg, 1) = (g^{-1}, 1)(x, 1)(g, 1) \in K.$$

Now $\langle (g^{-1}xg, 1) \mid g \in G \rangle$ is a subgroup of $K$ and $\langle g^{-1}xg \mid g \in G \rangle$ is a normal subgroup of $G$. Hence $\langle g^{-1}xg \mid g \in G \rangle = G$ (since $x \neq 1$) and so $G \times \{1\}$ is a subgroup of $K$. Now

$$\frac{G \times H}{G \times \{1\}} \simeq H$$

and $K$ corresponds to a normal subgroup of $H$ under the isomorphism. Hence either $K = G \times H$ or $K = G \times \{1\}$.

Similarly, if $(1,y) \in K$ where $y \neq 1$ then $\{1\} \times H$ is a subgroup of $K$ so $K = \{1\} \times H$ or $K = G \times H$.

The only other case to consider is when $K$ contains only $(1,1)$ and elements of the form $(x,y)$ where $x,y \neq 1$. But if $g \in G$ then there exists $h \in H$ with $(g,h) \in K$ since the mapping described by $(x,y) \mapsto (x,1)$ is a morphism whose image would be a proper non-trivial normal subgroup of $G$ if $g \in G$ did not appear in some element of $K$. Now either $G$ is abelian or there exist $g, x \in G$ with $g^{-1}xg \neq x$. Then $(g,h) \in K$ gives

$$(x^{-1}gx, h) = (x^{-1}, 1)(g, h)(x, 1) \in K.$$

Let $g' = x^{-1}gx \neq g$. Then $(g'^{-1}, h^{-1})(g,h) \in K$ gives $(g'^{-1}g, 1) \in K$, which is a contradiction.

Hence $G, H$ are abelian and so cyclic of prime order. Clearly, $|G| = |H|$ by question 1.21.

**1.23** Suppose that $G$ and $H$ are periodic. If $(g,h) \in G \times H$ then $g \in G$ so $g^n = 1$, and $h \in H$ so $h^m = 1$. Consequently we have that $(g,h)^{mn} = (1,1)$ and $G \times H$ is also periodic.

Suppose now that $G$ and $H$ are torsion-free. If $(g,h)^n = (1,1)$ then $g^n = 1$ and $h^n = 1$, which is a contradiction; hence $G \times H$ is also torsion-free.

**1.24** Suppose that $G = AB$ where $A, B$ are normal subgroups of $G$ with $A \cap B = N$. That $G/N \simeq A/N \times B/N$ follows from the following observations :

  (i) $A/N$ and $B/N$ are normal subgroups of $G/N$;

  (ii) If $gN \in G/N$ then $gN = aN.bN$ where $g = ab$;

  (iii) If $xN \in A/N \cap B/N$ then $xN = aN = bN$ gives $a^{-1}b \in N$ so that $b = an \in A$ and hence $b \in A \cap B = N$ and consequently $xN = bN = N$.

Consider $S_3$ and the subgroups $A = \langle(123)\rangle$ and $B = \langle(12)\rangle$. We have $S_3 = AB$ and $A \cap B = \{1\}$ but $S_3$ is not isomorphic to $C_3 \times C_2$.

**1.25** That $G \simeq A \times \operatorname{Ker} f$ follows from the following observations.

(a) $A$ and $\operatorname{Ker} f$ are normal subgroups of $G$.

(b) If $g \in G$ then $f(g) \in H$ so there exists $a \in A$ with $f(a) = f(g)$. Then $g = a(a^{-1}g)$ where $a \in A$ and $a^{-1}g \in \operatorname{Ker} f$.

(c) If $g \in A \cap \operatorname{Ker} f$ then $f_A(g) = f(g) = 1$ whence, since $f_A : A \to H$ is an isomorphism, $g = 1$.

The result is not true if $A$ is not normal in $G$. For example, consider the mapping $f : S_3 \to C_2 = \{1, a\}$ (where $a^2 = 1$) given by

$$f(x) = \begin{cases} 1 & \text{if } x = 1; \\ a & \text{if } x \text{ has order 2}; \\ 1 & \text{if } x \text{ has order 3}. \end{cases}$$

It is readily seen that $f$ is a morphism with $\operatorname{Ker} f = \langle(123)\rangle \simeq C_3$. The subgroup $A = \{(1), (12)\}$ of $S_3$ is not normal in $S_3$, and the restriction of $f$ to $A$ is an isomorphism onto $C_2$. However, $S_3$ is not isomorphic to $C_2 \times C_3$.

The isomorphisms stated in (i), (ii), (iii) and (iv) follow from the morphisms of question 1.12.

**1.26** Suppose that $C_2 \times C_2$ is generated by $x$ and by $y$. Then the subgroups of $C_2 \times C_2$ are

$$\{1\}, \langle x\rangle, \langle y\rangle, \langle xy\rangle, C_2 \times C_2.$$

The subgroup Hasse diagram is then

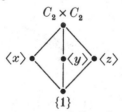

Suppose now that $G$ is a group and that the subgroup Hasse diagram of $G$ is

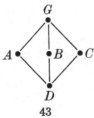

Clearly, $D = \{1\}$ and $A, B, C$ must be cyclic of prime order since they contain no proper subgroups other than 1. Let $A = \langle a \rangle$ and $B = \langle b \rangle$. We show first that $G \simeq A \times B$. For this purpose, we observe that $A$ and $B$ are normal in $G$. Suppose in fact that this were not the case. Then $\langle ab \rangle$ would not be the whole of $G$ (for otherwise $A$ and $B$ would be normal) and so we must have $\langle ab \rangle = C$ (since $\langle ab \rangle = A$ and $\langle ab \rangle = B$ lead to contradictions). Consider now $\langle b^{-1}ab \rangle$. This subgroup cannot be $G, B, C$ or $D$ and so it must be $A$. Similarly, $\langle a^{-1}ba \rangle = B$. Now $G$ must be generated by $a$ and $b$ (otherwise $\langle a, b \rangle$ would be properly contained in $G$). Hence $A$ and $B$ are normal in $G$, their intersection is $\{1\}$, and so $G = A \times B$.

It remains to show that $A \simeq B \simeq C_2$. For this purpose, consider $\langle ab^{-1} \rangle$. It is easy to see that this subgroup must be $C$, since the other possibilities lead to immediate contradictions. Now, since $G$ is abelian, we have $a^2 = abab^{-1} \in C$. But $a^2 \in A$, so $a^2 \in A \cap C = \{1\}$. In a similar way we have that $b^2 = 1$. Consequently, $G \simeq C_2 \times C_2$.

**1.27**   $C_2 \times C_2 \times C_2$ has 16 subgroups. Suppose that it is generated by $a, b, c$. Then the subgroup Hasse diagram is

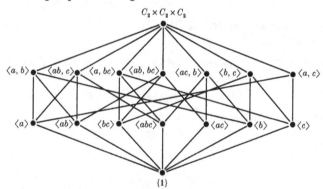

**1.28**   Consider the subgroups

$$H = \langle 2 \rangle = \{1, 2, 4, 8, 11, 16\} \simeq C_6 \,;$$
$$K = \langle 13 \rangle = \{1, 13\} \simeq C_2 \,.$$

We have that

(i)   $H$ and $K$ are normal subgroups of $G$;

(ii)   $G = HK$ (this is easily checked : for example, $5 = 2.13, 10 = 4.13$, etc.);

(iii)   $H \cap K = \{1\}$.

# Solutions to Chapter 1

Consequently, $G \simeq C_2 \times C_6$. This group is not cyclic since 2 is not coprime to 6.

The set of integers $n$ with $1 \leq n \leq 12$ and $n$ coprime to 12 is $\{1, 5, 7, 11\}$. Since

$$5^2 \equiv 1 \pmod{12}, \quad 7^2 \equiv 1 \pmod{12}, \quad 11^2 \equiv 1 \pmod{12},$$

we see that every non-trivial element has order 2. Hence the group is $C_2 \times C_2$ and is not cyclic.

**1.29** The only normal subgroups of $S_4$ (other than $\{1\}$ and $S_4$) are

$$V = \{(1), (12)(34), (13)(24), (14)(23)\}$$

and $A_4$, with $V \subset A_4 \subset S_4$.

The only normal subgroup of $S_5$ is $A_5$.

The only normal subgroup of $A_4$ is $V$ above.

The group $A_5$ is simple.

Consequently $S_4, S_5, A_4, A_5$ are indecomposable.

$\mathbb{R}^{\bullet} \simeq \mathbb{R}^{\bullet}_{>0} \times C_2$ (see question 1.25), and $C_6 \simeq C_2 \times C_3$.

$C_8$ is indecomposable. For, if $H$ and $K$ are subgroups of $C_8$ then either $H \subseteq K$ or $K \subseteq H$ (see question 1.8).

$\mathbb{C}^+ \simeq \mathbb{R}^+ \times \mathbb{R}^+$ (see question 1.25).

$\mathbb{Z}_{p^\infty}$ is indecomposable. For, if $H$ and $K$ are subgroups then either $H \subseteq K$ or $K \subseteq H$.

**1.30** If $xH$ has order $n$ then $x^n = h \in H$ so, by the hypothesis, there exists $h' \in H$ with $h'^n = h$. Now let $y = xh'^{-1}$. We have that $y^n = 1$. If $G/H$ is cyclic, take $xH$ as a generator and let $K = \langle y \rangle$. Then we have

(i) $H$ and $K$ are normal in $G$;

(ii) $yH$ generates $G/H$ and so, given $g \in G$, we have $g \in y^m H$ for some $m$, whence $g = y^m h$;

(iii) if $t \in H \cap K$ then $t = y^m$ for some $m < n$. But $y^m \notin H$ so we must have $t = 1$.

Consequently, $G \simeq H \times K$.

**1.31** Let $x$ have order $m$ and $y$ have order $n$. Then we have that $(xy)^{mn} = 1$, and so $xy$ has order at most $mn$. Suppose now that $z^{mn} = 1$ where $m$ and $n$ are coprime. Then there are integers $a$ and $b$ such that $am + bn = 1$, whence $z = z^{bn} z^{am} = xy$ where $x = z^{bn}$ and $y = z^{am}$. It follows that $x^m = z^{bmn} = 1$ and $y^n = z^{amn} = 1$. The orders of $x$ and $y$ are $m$ and $n$ respectively, for if their orders were less, say $m'$ and $n'$, then $z$ would be of order at most $m'n'$, a contradiction.

45

A similar argument extends this result to the case where $z$ is of order $m_1 m_2 \cdots m_k$ where $m_1, \ldots, m_k$ are pairwise coprime. In this case $m_1$ is coprime to $\prod_{i=2}^{k} m_i$ so, by the first part, we have $z = xy$ where $x$ has order $m_1$ and $y$ has order $\prod_{i=2}^{k} m_i$. The result now follows by induction on $k$.

That $G = H_1 \times H_2 \times \cdots \times H_k$ follows from the fact that

(i) each $H_i$ is a normal subgroup of $G$;

(ii) $G = H_1 H_2 \cdots H_k$ (by the above);

(iii) if $x \in H_i$ then $x^{p_i^{\alpha_i}} = 1$ which shows that $x$ does not belong to the product $H_1 \cdots H_{i-1} H_{i+1} \cdots H_k$.

If finally $r$ divides $|G|$ then $r$ is necessarily of the form

$$r = p_1^{\beta_1} p_2^{\beta_2} \cdots p_k^{\beta_k}$$

where $0 \leq \beta_i \leq \alpha_i$ for each $i$. If $K_i = \{x \in G \mid x^{p_i^{\beta_i}} = 1\}$, then $G$ has a subgroup of order $r$, namely the cartesian product of the subgroups $K_i$.

1.32  Let $I = \bigcap_{x \in G} x^{-1} H x$. If $t \in I$ then $t \in x^{-1} H x$ for every $x \in G$. Given $g \in G$ we then have $g^{-1} t g \in x^{-1} H x$ for $t \in (xg^{-1})^{-1} H x g^{-1}$. Hence $g^{-1} t g \in I$ and so $I$ is a normal subgroup of $G$.

If $A = \{g^{-1} x g \mid g \in G\}$ is a subgroup of $G$ then it must contain $1$ and so $g^{-1} x g = 1$ for some $g \in G$, which implies that $x = 1$. Thus we see that $A$ is a subgroup of $G$ if and only if $A = 1$, in which case it must be normal.

1.33  The subgroups $\{(1), (12)\}$ and $\{(1), (12)(34)\}$ of $S_4$ are not conjugate. All elements of order 3 in $S_4$ are conjugate, so all subgroups of order 3 are conjugate.

The elements $(123)$ and $(234)$ are not conjugate in $A_4$, for there is no $g \in A_4$ such that $g(123)g^{-1} = (234)$. In $S_4$ there are three such $g$, namely $(1234), (1324), (14)$.

1.34  Let $\{C_\lambda \mid \lambda \in \Lambda\}$ be the set of conjugacy classes of $G$. Suppose that $H$ is a subgroup which is a union of conjugacy classes, say $H = \bigcup_{\lambda \in \Lambda_1} C_\lambda$ where $\Lambda_1 \subseteq \Lambda$. If $h \in H$ then $h \in C_\lambda$ for some $\lambda \in \Lambda_1$ and so $g^{-1} h g \in C_\lambda \subseteq H$. Thus we see that $H$ is a normal subgroup of $G$. Conversely, if $H$ is a normal subgroup of $G$ then clearly every conjugate of $h \in H$ is contained in $H$ and so $H$ contains the conjugacy class of $G$ containing $h$. Thus $H$ is a union of conjugacy classes.

## Solutions to Chapter 1

In $S_4$ the conjugacy class of

| | |
|---|---|
| (1) | has 1 element |
| (12) | has 6 elements |
| (123) | has 8 elements |
| (1234) | has 6 elements |
| (12)(34) | has 3 elements. |

A normal subgroup is a union of conjugacy classes including the class $\{1\}$ with one element. By Lagrange's theorem, the order of a subgroup must therefore divide 24. The only possibilities are $1 + 3$ and $1 + 3 + 8$, so the only possible orders for non-trivial proper normal subgroups are 4 and 12.

The group $S_4$ has the normal subgroup $A_4$ with $|A_4| = 12$. Note that $A_4$ consists of all the even permutations and is the $1 + 3 + 8$ case above. The $1 + 3$ case gives the subgroup

$$\{(1), (12)(34), (13)(24), (14)(23)\}.$$

To see that this is a normal subgroup, it suffices to check that it is a subgroup; this follows easily from the fact that

$$(12)(34) \cdot (13)(24) = (14)(23), \quad \text{etc.}$$

**1.35**  The conjugacy classes are as follows.

| | | |
|---|---|---|
| (1) | 1 element of order 1 | even |
| (12) | 10 elements of order 2 | odd |
| (123) | 20 elements of order 3 | even |
| (1234) | 30 elements of order 4 | odd |
| (12345) | 24 elements of order 5 | even |
| (12)(34) | 15 elements of order 2 | even |
| (12)(345) | 20 elements of order 6 | odd |

The only normal subgroup of $S_5$ is $A_5$ (use the method of the previous question). The conjugacy classes of $A_5$ are the classes marked 'even' above, with the exception that the 24 elements of order 5 break into two classes of 12 elements, one containing (12345) and the other (13524).

47

**1.36** The first statement is a consequence of the observation that two conjugates $x^{-1}ax$ and $y^{-1}ay$ are equal if and only if $(xy^{-1})^{-1}axy^{-1} = a$, which is equivalent to $xy^{-1} \in \mathcal{N}_G(a)$, which is the case if and only if $\mathcal{N}_G(a)y = \mathcal{N}_G(a)x$.

Now the conjugacy class of a cycle of length $n$ in $S_n$ consists of all the cycles of order $n$. Thus it contains $(n-1)!$ elements. For a given cycle $a$, the index of $\mathcal{N}_G(a)$ in $S_n$ is therefore $(n-1)!$. Since $|S_n| = n!$ it follows that $|\mathcal{N}_G(a)| = n$. Since $a$ has $n$ distinct powers which commute with it, no other elements of $S_n$ can commute with $a$.

Suppose now that $n$ is odd with $n \geq 3$. Cycles of length $n$ are even permutations, so are in $A_n$. Suppose that $a$ is a cycle of length $n$. Only the $n$ powers of $a$ are in $\mathcal{N}_{S_n}(a)$ (by the above argument) and so the conjugacy class of $a$ contains $\frac{1}{2}n!/n = \frac{1}{2}(n-1)!$ elements. Since there are $(n-1)!$ cycles of length $n$, there must be two conjugacy classes each containing $\frac{1}{2}(n-1)!$ elements.

Suppose now that $n$ is even with $n \geq 4$. Then $n-1$ is odd, so a cycle of length $n-1$ is even and therefore belongs to $A_n$. There are $n(n-2)!$ cycles of length $n-1$. Let $a$ be such a cycle. Then in $S_n$ the conjugacy class of $a$ contains $n(n-2)!$ elements and so $|\mathcal{N}_{S_n}(a)| = n-1$. But if $x \in \mathcal{N}_{S_n}(a)$ then, since $n-1$ is odd, $x$ is an even permutation, so $x \in A_n$. Thus we have that $\mathcal{N}_{S_n}(a) = \mathcal{N}_{A_n}(a)$, so the conjugacy class of $a$ in $A_n$ contains $\frac{1}{2}n!/(n-1) = \frac{1}{2}n(n-2)!$ elements. Thus there are two conjugacy classes of cycles of length $n-1$ in $A_n$ each containing $\frac{1}{2}n(n-2)!$ elements.

**1.37** $x \in G$ commutes with $a \in G$ if and only if it commutes with $a^{-1}$. Hence $\mathcal{N}_G(a) = \mathcal{N}_G(a^{-1})$. The number of conjugates of $a$, being the index of $\mathcal{N}_G(a)$ in $G$, must therefore be the same as the number of conjugates of $a^{-1}$.

Suppose now that $|G|$ is even and that 1 is the only element of $G$ that is conjugate to its inverse. For each conjugacy class $A_i$ let $B_i$ be that containing the inverses. Then we have

$$G = \{1\} \cup A_1 \cup \ldots \cup A_k \cup B_1 \cup \ldots \cup B_k$$

from which we obtain

$$|G| = 1 + |A_1| + \ldots + |A_k| + |B_1| + \ldots + |B_k|$$
$$= 1 + 2|A_1| + \ldots + 2|A_k|$$

since $|B_i| = |A_i|$ for each $i$. This contradicts the fact that $|G|$ is even. Thus we conclude that there is at least one element $a \neq 1$ with $a$ conjugate to $a^{-1}$.

## Solutions to Chapter 1

**1.38** Consider $D_{2n}$ with generators $a$ and $b$ such that $a^2 = b^n = 1$ and $aba = b^{-1}$. The elements of $D_{2n}$ are

$$\{1, b, b^2, \ldots, b^{n-1}, a, ab, ab^2, \ldots, ab^{n-1}\}.$$

Now since

$$b^{-1}(ab^i)b = b^{-1}ab^{i+1} = ab.b^{i+1} = ab^{i+2}$$

we see that $ab^i$ is conjugate to $ab^{i+2}$ for every $i$. Also, since $ab^i a = b^{-i}$ we see that $b^i$ is conjugate to $b^{-i}$.

Suppose that $n$ is odd. Then the conjugacy classes are

$$\{1\}, \quad \{a, ab, \ldots, ab^{n-1}\}, \quad \{b^i, b^{-i}\}$$

where $1 \leq i \leq \frac{1}{2}(n-1)$.

If $n$ is even, the conjugacy classes are

$$\{1\}, \quad \{a, ab^2, \ldots, ab^{n-2}\}, \quad \{ab, ab^3, \ldots, ab^{n-1}\}, \quad \{b^{\frac{1}{2}n}\}, \quad \{b^i, b^{-i}\}$$

where $1 \leq i \leq \frac{1}{2}(n-2)$.

**1.39** Suppose that $K = x^{-1}Hx$. We show that $N_G(K) = x^{-1}N_G(H)x$. For this purpose, let $a \in N_G(K)$. To show that $a \in x^{-1}N_G(H)x$ we must show that $xax^{-1} \in N_G(H)$. So let $h \in H$ and consider

$$(xax^{-1})^{-1}hxax^{-1} = xa^{-1}x^{-1}hxax^{-1}.$$

Now $x^{-1}hx = k \in K$ since $x^{-1}Hx = K$, and $a^{-1}ka = k' \in K$ since $a \in N_G(K)$. Hence

$$(xax^{-1})^{-1}hxax^{-1} = xk'x^{-1} = h' \in H$$

since $xKx^{-1} = H$. Thus $xax^{-1} \in N_G(H)$ as required.

**1.40** Let $H = \{1, h\}$. For every $g \in G$ we have $g^{-1}hg \in H$, so $g^{-1}hg = 1$ or $g^{-1}hg = h$. The former gives the contradiction $h = 1$. Hence $g^{-1}hg = h$ and $H \subseteq Z(G)$.

That $H$ is not necessarily a subgroup of the derived group of $G$ can be seen by taking $G = C_2$. Here we have that $C_2$ is a normal subgroup of $C_2$ but the derived group of $C_2$ is $\{1\}$.

Suppose that $x$ is the only element of order 2 in $G$. Let $g \in G$ and consider $g^{-1}xg = y$ say. We have $y^2 = g^{-1}x^2g = 1$ and certainly $g^{-1}xg \neq 1$ (since otherwise $x = 1$). Thus $y$ is an element of order 2 and so, by the hypothesis, we have $y = x$. Consequently $g^{-1}xg = x$ and $x \in Z(G)$.

*1.41*   Let $n \in N$ and $g \in G$. Then $[n,g] = n^{-1}g^{-1}ng \in G'$. Since $N$ is a normal subgroup of $G$ we have $n^{-1}(g^{-1}ng) \in N$. Consequently $[n,g] \in N \cap G' = 1$. It follows that $[n,g] = 1$ and hence that $n \in Z(G)$.

   Clearly, if $z \in Z(G)$ then

$$zN.gN = zgN = gzN = gN.zN$$

and so $Z(G)/N \subseteq Z(G/N)$. To obtain the converse inclusion we observe that

$$
\begin{aligned}
zN \in Z(G/N) &\Longrightarrow (\forall g \in G)\ zN.gN = gN.zN \\
&\Longrightarrow (\forall g \in G)\ z^{-1}g^{-1}zgN = N \\
&\Longrightarrow (\forall g \in G)\ [g,z] \in N \\
&\Longrightarrow (\forall g \in G)\ [g,z] \in N \cap G' = 1 \\
&\Longrightarrow z \in Z(G).
\end{aligned}
$$

# Solutions to Chapter 2

**2.1**  Let $|G| = p^\alpha$. The class equation gives

$$|G| = |Z(G)| + \sum_{g'_\lambda} |G : N(g'_\lambda)|$$

where $g'_\lambda \in C_\lambda$ for $|C_\lambda| > 1$. Now $p$ divides both $|G|$ and $|G : N(g'_\lambda)|$, so $p$ divides $|Z(G)|$. Hence $Z(G)$ is non-trivial.

Now suppose that $|G| = p^2$. Since $Z(G)$ is non-trivial we have $|Z(G)| = p$ or $|Z(G)| = p^2$. Now if $|Z(G)| = p$ then $|G/Z(G)| = p$ so is cyclic. Let $aZ(G)$ be a generator of $G/Z(G)$. Then two arbitrary elements of $G$ are $a^m x$ and $a^n y$ where $x, y \in Z(G)$. But $a^m x a^n y = a^n y a^m x$ since $x, y \in Z(G)$. Hence $G$ is abelian, and so $|G| = |Z(G)| = p$ which is a contradiction. It follows that we must have $|Z(G)| = p^2 = |G|$ whence $G$ is abelian.

The groups of order 9 are $C_3 \times C_3$ and $C_9$.

**2.2**  If $x \in A \cap B$ then $\vartheta(x) \in \vartheta(A)$ and $\vartheta(x) \in \vartheta(B)$ and consequently we see that $\vartheta(A \cap B) \le \vartheta(A) \cap \vartheta(B)$. Now $\vartheta^{-1}$ is an automorphism and so we also have that $\vartheta^{-1}(X \cap Y) \le \vartheta^{-1}(X) \cap \vartheta^{-1}(Y)$. Now put $X = \vartheta(A)$ and $Y = \vartheta(B)$ to get $\vartheta^{-1}[\vartheta(A) \cap \vartheta(B)] \le A \cap B$, whence $\vartheta(A) \cap \vartheta(B) \le \vartheta(A \cap B)$. Hence we have the equality $\vartheta(A \cap B) = \vartheta(A) \cap \vartheta(B)$.

**2.3**  Let $\varphi_x \in \operatorname{Inn} G$ be the automorphism $\varphi_x : g \mapsto x^{-1} g x$. Let $\vartheta \in \operatorname{Aut} G$; we have to show that $\vartheta^{-1} \varphi_x \vartheta \in \operatorname{Inn} G$. Now

$$\vartheta^{-1} \varphi_x \vartheta : g \mapsto \vartheta^{-1}(x^{-1} \vartheta(g) x) = \vartheta^{-1}(x^{-1})\, g\, \vartheta^{-1}(x).$$

But if $\vartheta^{-1}(x) = y$ then $\vartheta^{-1}(x^{-1}) = y^{-1}$ and we have

$$\vartheta^{-1} \varphi_x \vartheta : g \mapsto y^{-1} g y,$$

so $\vartheta^{-1}\varphi_x\vartheta \in \operatorname{Inn} G$. To show that $G/Z(G) \simeq \operatorname{Inn} G$ define $\psi : G \to \operatorname{Inn} G$ by $\psi(x) = \varphi_{x^{-1}} : g \mapsto xgx^{-1}$. Then

$$\psi(xy) = \varphi_{(xy)^{-1}} = \varphi_{y^{-1}x^{-1}} : g \mapsto xygy^{-1}x^{-1}$$
$$= \psi(x)\psi(y).$$

Clearly, $\psi$ is surjective, and

$$x \in \operatorname{Ker}\psi \implies (\forall g \in G)\ xgx^{-1} = g \implies x \in Z(G).$$

Thus $\operatorname{Ker}\psi = Z(G)$ and $\operatorname{Inn} G = \operatorname{Im}\psi \simeq G/\operatorname{Ker}\psi = G/Z(G)$.

Clearly, $b^2 \in Z(D_8)$ since $a^{-1}b^2a = (a^{-1}ba)^2 = (b^{-1})^2 = b^2$. Now we have $D_8 = \{1, b, b^2, b^{-1}, a, ab, ab^2, ab^{-1}\}$ and $a, b \notin Z(D_8)$ since otherwise $D_8$ would be abelian. Hence $b^{-1} \notin Z(D_8)$. Since $b^2 \in Z(D_8)$ and $a \notin Z(D_8)$ we have $ab^2 \notin Z(D_8)$. Also $ab, ab^{-1}$ commute with $a$ if and only if $b$ commutes with $a$, so $ab, ab^{-1} \notin Z(D_8)$. Hence $Z(D_8) = \langle b^2 \rangle \simeq C_2$.

$$D_8/Z(D_8) \simeq \langle a, b \mid a^2 = 1,\ b^4 = 1,\ a^{-1}ba = b^{-1},\ b^2 = 1 \rangle$$
$$= \langle a, b \mid a^2 = 1,\ b^2 = 1,\ ab = ba \rangle$$
$$\simeq C_2 \times C_2.$$

Similarly, $Z(Q_8) = \langle x^2 \rangle \simeq C_2$ and $Q_8/Z(Q_8) \simeq C_2 \times C_2$.

2.4    Both questions have a negative answer. The same group, namely the dihedral group of order 8, serves to provide counter-examples.

Let $D_8 = \langle a, b \mid a^2 = 1,\ b^4 = 1,\ (ab)^2 = 1 \rangle$. If $D_8$ were the direct product of two non-trivial subgroups, one must be of order 4 and the other of order 2. But a normal subgroup of order 2 is central, so $D_8 = A \times B$ where $|A| = 4$ and $B = Z(D_8) = \langle b^2 \rangle$. Now $b \notin A$ since $A \cap B = \{1\}$, so $Ab \in D_8/A \simeq C_2$. Hence $(Ab)^2 = Ab^2 = A$, showing that $b^2 \in A$. Thus $B \subseteq A$, which contradicts $A \cap B = \{1\}$. Thus $D_8$ is indecomposable.

However, there is a subgroup and a quotient group of $D_8$ each of which is isomorphic to $C_2 \times C_2$; for the subgroup, take $\langle a, b^2 \rangle$ (or $\langle ab, b^2 \rangle$), and note that $D_8/Z(D_8) \simeq C_2 \times C_2$ since $Z(D_8) = \langle b^2 \rangle$.

2.5    Since $G/Z(G)$ is cyclic, every element is a power of a single element, say $aZ(G)$. Thus, given $g \in G$, we can write $g = a^n z$ for some $z \in Z(G)$ and $n \in \mathbb{Z}$. If $g_1, g_2 \in G$ then, with an obvious notation, we have

$$g_1 g_2 = a^{n_1} z_1 a^{n_2} z_2 = a^{n_1 + n_2} z_1 z_2$$

since $z_2$ is central; and

$$g_2 g_1 = a^{n_2} z_2 a^{n_1} z_1 = a^{n_1 + n_2} z_1 z_2$$

since $z_1, z_2$ are central. Thus $G$ is abelian.

Suppose that $\operatorname{Aut} G$ is cyclic. Then $\operatorname{Inn} G$ is cyclic. But $G/Z(G) \simeq \operatorname{Inn} G$, so $G/Z(G)$ is cyclic and hence $G$ is abelian.

**2.6**   If $\alpha \in \operatorname{Aut} S_3$ then $\alpha$ must map an element of order 2 in $S_3$ to an element of order 2. Hence $\alpha$ permutes the set $A = \{(12), (23), (13)\}$. If $\alpha$ leaves all three of these elements fixed then, since the elements of order 2 generate $S_3$, $\alpha$ must be the identity map. Hence if $\alpha, \beta \in \operatorname{Aut} S_3$ give the same permutation of $A$ then $\alpha\beta^{-1}$ leaves the elements of $A$ fixed, whence $\alpha\beta^{-1} = \mathrm{id}$ and $\alpha = \beta$. Hence $|\operatorname{Aut} S_3| \leq 6$.

Now since $Z(S_3) = \{1\}$ and $S_3/Z(S_3) \simeq \operatorname{Inn} S_3$ it follows that $S_3$ has six inner automorphisms. Then

$$6 = |\operatorname{Inn} S_3| \leq |\operatorname{Aut} S_3| \leq 6$$

gives $\operatorname{Inn} S_3 = \operatorname{Aut} S_3$, so that $\operatorname{Aut} S_3 \simeq S_3/Z(S_3) = S_3$ as required.

**2.7**   Let $C_2 \times C_2 = \langle a, b \mid a^2 = b^2 = a^{-1}b^{-1}ab = 1 \rangle$. Then we have $C_2 \times C_2 = \{1, a, b, ab\}$ and each element in the set $\{a, b, ab\}$ is of order 2. If $\alpha \in \operatorname{Aut}(C_2 \times C_2)$ then $\alpha$ fixes 1 and permutes $a, b, ab$. Thus $\alpha$ is completely determined by its action on $\{a, b, ab\}$. Hence $\operatorname{Aut}(C_2 \times C_2) \leq S_3$. To show that $\operatorname{Aut}(C_2 \times C_2) = S_3$, it remains to show that every permutation on $\{a, b, ab\}$ gives an automorphism of $C_2 \times C_2$. This follows easily on noting that the product of any two distinct elements of $\{a, b, ab\}$ yields the third element, and this property is preserved under a bijection.

Now, as shown in question 2.6, $\operatorname{Aut} S_3 = S_3$. Hence we have that $C_2 \times C_2$ and $S_3$ have isomorphic automorphism groups.

**2.8**   If $\vartheta \in Z(\operatorname{Aut} G)$ and $\varphi_g$ is the inner automorphism given by $\varphi_g(x) = g^{-1}xg$ then we must have $\vartheta\varphi_g = \varphi_g\vartheta$. Hence, for every $x \in G$, we have

$$\vartheta(g^{-1})\vartheta(x)\vartheta(g) = \vartheta(g^{-1}xg) = g^{-1}\vartheta(x)g$$

and so $g\vartheta(g^{-1})\vartheta(x) = \vartheta(x)g\vartheta(g^{-1})$. Since $\vartheta$ is surjective, we see that $g\vartheta(g^{-1}) \in Z(G) = \{1\}$ and therefore $\vartheta(g) = g$. Since this holds for all $g \in G$ we have that $\vartheta = \mathrm{id}$ as required.

To see that the converse is false, note that $S_3 = \operatorname{Aut}(C_2 \times C_2)$ (by question 2.7) and $Z(S_3) = \{1\}$, yet $Z(C_2 \times C_2) = C_2 \times C_2 \neq \{1\}$.

**2.9**   It is clear that the additive group of the vector space $\mathbf{Z}_p^n$ is isomorphic to $C_p \times C_p \times \cdots \times C_p$ (with $n$ terms). Suppose that $\vartheta$ is an automorphism of the additive group of $\mathbf{Z}_p^n$. To see that $\vartheta$ is a linear mapping on $\mathbf{Z}_p^n$ it suffices to observe that

$$\vartheta[(a_1, \ldots, a_n) + (b_1, \ldots, b_n)] = \vartheta(a_1, \ldots, a_n) + \vartheta(b_1, \ldots, b_n)$$

gives

$$\vartheta[m(a_1, \ldots, a_n)] = \vartheta(ma_1, \ldots, ma_n)$$
$$= m\vartheta(a_1, \ldots, a_n).$$

Thus $\vartheta$ is an invertible linear map. This shows that $\mathrm{Aut}\,G$ is isomorphic to the group of invertible linear maps on the vector space $\mathbf{Z}_p^n$. Now fix a basis $B$ of $\mathbf{Z}_p^n$. Then the mapping that associates with each invertible linear map on $\mathbf{Z}_p^n$ its $n \times n$ matrix relative to $B$ is clearly an isomorphism onto $\mathrm{GL}(n, p)$.

**2.10**   If $\mathrm{Aut}\,G = \{1\}$ then $G$ must be abelian. For, given $g \in G$, the inner automorphism of conjugation by $g$ is trivial if and only if $g \in Z(G)$. Hence $G = Z(G)$ and so is abelian.

Suppose now that $G$ contains an element $g$ of order greater than 2. Since $G$ is abelian, the mapping described by $\vartheta : x \mapsto x^{-1}$ is a group morphism. Since $\vartheta(g) = g^{-1} \neq g$ we see that $\vartheta$ is a non-trivial element of $\mathrm{Aut}\,G$, a contradiction. Thus every element of $G$ must have order 2.

It now follows that $G$ is a vector space over $\mathbf{Z}_2$. If the dimension of this vector space is greater than 1 then every non-trivial permutation of the basis elements induces a non-trivial automorphism on $G$. Hence the dimension is at most 1. Consequently we have that either $G \simeq C_2$ or $G$ is trivial.

**2.11**   (a) True. Let $\vartheta : G \to G$ be a group morphism. Then for $a, b \in G$ we have

$$\vartheta([a, b]) = \vartheta(a^{-1}b^{-1}ab) = \vartheta(a)^{-1}\vartheta(b)^{-1}\vartheta(a)\vartheta(b)$$
$$= [\vartheta(a), \vartheta(b)] \in G'.$$

Hence $\vartheta(G') \subseteq G'$.

(b) False. Consider $C_2 \times S_3$ where $C_2 = \langle a \rangle$. $Z(C_2 \times S_3)$ is the subgroup $\langle (a, 1) \rangle$. Consider the mapping $\vartheta : C_2 \times S_3 \to C_2 \times S_3$ given by setting

$$(\forall \pi \in S_3) \qquad \begin{cases} \vartheta(a, \pi) = (1, (12)), \\ \vartheta(1, \pi) = (1, 1). \end{cases}$$

Then $\vartheta$ is a group morphism but $\vartheta(Z(C_2 \times S_3)) \not\subseteq Z(C_2 \times S_3)$.

(c) False. $A_4$ contains only one non-trivial proper normal subgroup $V \simeq C_2 \times C_2$. Suppose that $\vartheta : A_4 \to A_4$ is a group morphism. Then $\operatorname{Ker}\vartheta$ must be $\{1\}, V$ or $A_4$ and it follows from this that $V$ is fully invariant.

(d) True. If $\vartheta : G \to G$ is a group morphism then from $\vartheta(g^n) = [\vartheta(g)]^n$ we obtain $\vartheta(G^n) \subseteq G^n$.

(e) True. If $\vartheta : G \to G$ is a group morphism then

$$g^n = 1 \Longrightarrow \vartheta(g^n) = 1 \Longrightarrow [\vartheta(g)]^n = 1$$

gives $\vartheta(G_n) \subseteq G_n$.

**2.12** It suffices to show that $x$ is conjugate to $y$ if and only if $\alpha(x)$ is conjugate to $\alpha(y)$. But

$$x = g^{-1}yg \iff \alpha(x) = \alpha(g^{-1}yg) = [\alpha(g)]^{-1}\alpha(y)\alpha(g)$$

so the result follows.

To show that $N$ is normal in $\operatorname{Aut}G$ we must show that if $\alpha \in N, \beta \in \operatorname{Aut}G$ then $(\beta^{-1}\alpha\beta)(C) = C$. But $\beta(C)$ is a conjugacy class by the first part of the question, so $\alpha[\beta(C)] = \beta(C)$ by the definition of $\alpha$. Hence $(\beta^{-1}\alpha\beta)(C) = \beta^{-1}[\beta(C)] = C$ as required.

**2.13** To show that $NC \triangleleft G$ it suffices to show that $C \triangleleft G$ since $N$ is given to be normal. So let $g \in G$ and $c \in C$. Given $n \in N$ there exists $n' \in N$ such that $gn = n'g$ and so, since $C$ centralises $N$,

$$g^{-1}cgn = g^{-1}cn'g = g^{-1}n'cg = ng^{-1}cg.$$

For $g \in G$ we have $\psi_g \in A$ where $\psi_g : n \mapsto gng^{-1} \in N$. Define $\psi : G \to A$ by $\psi(G) = \psi_g$. Note that $\operatorname{Ker}\psi = C$. We show that $\psi(NC) \subseteq I$, whence $\psi$ induces a morphism from $G/NC$ to $A/I$. Now $\psi(C) \subseteq I$ since $C = \operatorname{Ker}\psi$; and if $n \in N$ then $\psi(n) = \psi_n \in I$ so $\psi(N) \subseteq I$ whence $\psi(NC) \subseteq I$.

To show that the mapping from $G/NC$ to $A/I$ induced by $\psi$ is injective, we must show that $\{g \in G \mid \psi(g) \in I\} = NC$. But this is clear from the fact that $\psi(N) = I$ and $\operatorname{Ker}\psi = C$.

Also, since $N/Z(N) \simeq I$ and $Z(N) = N \cap C$ we have

$$I \simeq N/(N \cap C) \simeq NC/C.$$

If $Z(N) = \{1\}$ then $N \cap C = \{1\}$. Also, since $A/I$ is trivial so is $G/NC$. Hence $G = NC$ and we have shown that $G = N \times C$.

Since $Z(S_3) = \{1\}$ we have $\operatorname{Inn}S_3 \simeq S_3/Z(S_3) \simeq S_3$. But $\operatorname{Aut}S_3 = S_3$ (see question 2.6), so every automorphism of $S_3$ is inner. Thus $S_3$ satisfies the conditions required of the subgroup $N$ and the result follows.

**2.14**   (a) If $\vartheta(H) \leq H$ and $\vartheta^{-1}(H) \leq H$ then from the latter we have $H \leq \vartheta(H)$ whence equality follows.

(b) If $x \in \bigcap H_\lambda$ where each $H_\lambda$ is characteristic then for every $\vartheta \in \operatorname{Aut} G$ we have $\vartheta(x) \in H_\lambda$ for all $\lambda$, whence $\vartheta(x) \in \bigcap H_\lambda$.

(c) If $x \in HK$ then $x = hk$ gives $\vartheta(x) = \vartheta(h)\vartheta(k) \in HK$.

(d) Let $c \in C = [H, K]$. Then $c = t_1 t_2 \cdots t_n$ where $t_i \in [h_i, k_i]^{\epsilon_i}$ with $\epsilon_i = \pm 1$. Then $\vartheta(c) = \vartheta(t_1) \cdots \vartheta(t_n)$ with $\vartheta(t_i) = \vartheta[h_i, k_i]^{\epsilon_i} = [\vartheta(h_i), \vartheta(k_i)]^{\epsilon_i}$.

(e) $H$ is normal in $G$, so if $\varphi$ is an inner automorphism of $G$ then $\varphi(H) \leq H$. But if $\varphi_H$ is the restriction of $\varphi$ to $H$ we have $\varphi_H(K) \leq K$, for certainly $\varphi_H$ is an automorphism (not necessarily inner) and $K$ is characteristic in $H$. But $\varphi(K) = \varphi_H(K)$ so $\varphi(K) \leq K$ gives $K \triangleleft G$.

**2.15**   Let $\vartheta \in \operatorname{Aut} G$ and let $K = \vartheta(H)$. We have to show that $K \subseteq H$. Suppose that $|H| = n$ and $|G/H| = m$. Since $HK/H \leq G/H$ we have that $|HK/H|$ divides $m$. But we know that $HK/H \simeq K/(H \cap K)$, so $|HK/H|$ divides $n$. But h.c.f.$(n, m) = 1$ by hypothesis. Hence $|HK/H| = 1$, so $HK = H$ and $K \leq H$ as required.

**2.16**   If $a \in F$ then $a^{-1} \in F$ since $(g^{-1}ag)^{-1} = g^{-1}a^{-1}g$. But if $a, b \in F$ we have $ab \in F$, for $g^{-1}abg = g^{-1}agg^{-1}bg$. Hence $F \leq G$. Now $F$ is characteristic in $G$. To see this, let $a \in F$ and $\vartheta \in \operatorname{Aut} G$. Writing $g = \vartheta(g')$ we have $g^{-1}\vartheta(a)g = \vartheta(g'^{-1}ag')$ and so there are only finitely many conjugates of $\vartheta(a)$. Thus $\vartheta(a) \in F$.

**2.17**   $\vartheta_t$ is clearly an additive group morphism which, since $t \neq 0$, is injective. Since $\vartheta_t(t^{-1}x) = x$ we see that $\vartheta_t$ is also surjective. Hence $\vartheta_t$ is an automorphism.

Let $H$ be a non-trivial characteristic subgroup of $\mathbb{Q}^+$. For every $t \in \mathbb{Q} \setminus \{0\}$ we have $\vartheta_t(H) = H$. We show as follows that $H = \mathbb{Q}^+$. Since $H$ is non-trivial, choose $x \in H$ with $x \neq 0$ and let $y \in \mathbb{Q} \setminus \{0\}$. Then for $t = yx^{-1} \neq 0$ we have

$$y = tx = \vartheta_t(x) \in \vartheta_t(H) = H,$$

from which $H = \mathbb{Q}^+$ follows.

**2.18**   A Sylow $p$-subgroup of $H$ is a $p$-subgroup of $G$ that is contained in a Sylow $p$-subgroup. Let $P_1$ and $P_2$ be Sylow $p$-subgroups of $H$. Let $P_1 \leq P$ and $P_2 \leq P$ where $P$ is a Sylow $p$-subgroup of $G$. Then $H \cap P$ is a $p$-subgroup of $H$ and $P_1 \leq H \cap P, P_2 \leq H \cap P$ implies $P_1 = H \cap P = P_2$.

Now suppose that $H \triangleleft G$ and that $P$ is a Sylow $p$-subgroup of $G$. Then $H \cap P$ is a $p$-subgroup of $H$ so is contained in a Sylow $p$-subgroup $P_1$ of

$H$. Now $P_1 \leq \overline{P}$ where $\overline{P}$ is a Sylow $p$–subgroup of $G$. Then $\overline{P} = g^{-1}Pg$ for some $g \in G$ since Sylow $p$–subgroups of $G$ are conjugate. Now since $H \triangleleft G$ we have

$$P_1 = \overline{P} \cap H = g^{-1}Pg \cap H = g^{-1}(P \cap H)g.$$

Hence $|P \cap H| = |P_1|$ and, since $P \cap H \leq P_1$, it follows that $P \cap H = P_1$. Let $|G| = p^n k$. Then $|P| = p^n$. Let $|H| = p^m t$. Then $|H \cap P| = p^m$. Now $|G/H| = p^{n-m}s$ where $st = k$ and

$$|HP/H| = |P|/|P \cap H| = p^{n-m}.$$

Hence $HP/H$ is a Sylow $p$–subgroup of $G/H$.

Suppose now that we drop the condition that $H$ be normal in $G$. Consider $G = S_3$ and $H = \langle (12) \rangle$. We have that $P = \langle (13) \rangle$ is a Sylow 2–subgroup of $S_3$, but $|H| = 2$ and $H \cap P = \{1\}$.

**2.19**  Let $H$ be a normal $p$–subgroup of $G$. Then $H \leq P$ where $P$ is a Sylow $p$–subgroup of $G$. But every Sylow $p$–subgroup of $G$ is of the form $g^{-1}Pg$ for some $g \in G$, and $H \leq P$ implies $H = g^{-1}Hg \leq g^{-1}Pg$ since $H \triangleleft G$.

Let $H_1, \ldots, H_n$ be distinct normal Sylow $p$–subgroups of $G$. Since a normal Sylow $p$–subgroup is unique, every element of $p$–power order is contained in this Sylow $p$–subgroup. Now observe that

(a) $H_1, \ldots, H_n \triangleleft G$ by hypothesis.
(b) $G = H_1 \cdots H_n$. This follows from the fact that $H_i \cap H_j = \{1\}$ for $i \neq j$ and $|H_1 \cdots H_n| = |H_1| \cdots |H_n| = |G|$.
(c) $H_i \cap H_1 \cdots H_{i-1}H_{i+1} \cdots H_n = \{1\}$. This follows from the fact that if $a \in H_i$ and $b \in H_j$ with $i \neq j$ then $[a,b] \in H_i \cap H_j = \{1\}$.

Thus we see that $G$ is the direct product of its Sylow $p$–subgroups.

**2.20**  We have that $|A_5| = 60 = 5 \cdot 3 \cdot 2^2$. Hence the Sylow 5–subgroups are $C_5$, the Sylow 3–subgroups are $C_3$, and the Sylow 2–subgroups are $C_2 \times C_2$ since $A_5$ has no element of order 4.

There are $1 + 5k$ Sylow 5–subgroups where $1 + 5k$ divides 60; thus there are six Sylow 5–subgroups. There are $1 + 3k$ Sylow 3–subgroups where $1 + 3k$ divides 60; thus there are four or ten Sylow 3–subgroups. There are in fact ten Sylow 3–subgroups as a little computation will show. There are $1 + 2k$ Sylow 2–subgroups where $1 + 2k$ divides 60; thus there are three, five or fifteen Sylow 2–subgroups. There are in fact five Sylow 2–subgroups as a little computation will show.

**2.21**   Let $g \in G$. Then $g^{-1}Pg \leq g^{-1}Kg = K$ since $K \lhd G$. Thus, since $|g^{-1}Pg| = |P|$, we have that $g^{-1}Pg$ is also a Sylow $p$–subgroup of $K$. Hence, by Sylow's theorem, $P$ and $g^{-1}Pg$ are conjugate in $K$. It follows that $g^{-1}Pg = k^{-1}Pk$ for some $k \in K$. Then $(gk^{-1})^{-1}Pgk^{-1} = P$ and so $gk^{-1} \in \mathcal{N}(P)$ whence $g \in \mathcal{N}(P)K$. Since this holds for all $g \in G$, we have that $G = \mathcal{N}(P)K$.

Let $\mathcal{N}(\overline{P}) \leq H \leq G$. Since $\overline{P} \leq H \leq G$, we have that $\overline{P}$ is a Sylow $p$–subgroup of $H$. Let $L = \mathcal{N}(H)$. By the first part of the question, $L = \mathcal{N}_L(\overline{P})H$ where $\mathcal{N}_L(\overline{P})$ is the normaliser of $\overline{P}$ in $L$. But $\mathcal{N}_L(\overline{P}) \leq \mathcal{N}(\overline{P}) \leq H$, so $L = H$ as required.

**2.22**   Let $S \leq G$ and suppose that $P$ is a Sylow $p$–subgroup of $S$. We have to show that $P$ is cyclic. Now $P$ is a $p$–subgroup of $G$ so $P \leq \overline{P}$ for some Sylow $p$–subgroup $\overline{P}$ of $G$. Since $\overline{P}$ is cyclic, so then is $P$.

Suppose that $P_1$ and $P_2$ are $p$–subgroups of $G$ with $|P_1| = |P_2|$. We have $P_1 \leq \overline{P_1}$ and $P_2 \leq \overline{P_2}$ where $\overline{P_1}, \overline{P_2}$ are Sylow $p$–subgroups of $G$. But $\overline{P_1}$ is conjugate to $\overline{P_2}$, so there exists $g \in G$ with $g^{-1}\overline{P_1}g = \overline{P_2}$. But now $g^{-1}P_1g \leq \overline{P_2}$ and $|g^{-1}P_1g| = |P_1| = |P_2|$, so $g^{-1}P_1g$ and $P_2$ are subgroups of the same order in the cyclic group $\overline{P_2}$. Hence $g^{-1}P_1g = P_2$ as required.

Certainly $|N \cap H|$ divides $|N|$ and $|H|$ and so divides h.c.f.$(|N|, |H|)$. Let $p$ be a prime divisor of h.c.f.$(|N|, |H|)$ and suppose that $p^n$ is the highest power of $p$ that divides it. Then $p^n$ divides $|N|$ and $|H|$ and so there exist $P_1, P_2$ with $P_1 \leq N, P_2 \leq H$ and $|P_1| = p^n = |P_2|$. Now $P_1$ is conjugate to $P_2$, and since $N \lhd G$ we must have $P_2 \leq N$. Hence $P_2 \leq H \cap N$ and $|H \cap N|$ is divisible by $p^n$. Thus

$$|H \cap N| = \text{h.c.f.}(|H|, |N|)$$

and from the isomorphism $HN/N \simeq H/(H \cap N)$ we obtain

$$|HN| = \frac{|H|\,|N|}{|H \cap N|} = \frac{|H|\,|N|}{\text{h.c.f.}(|H|, |N|)} = \text{l.c.m.}(|H|, |N|).$$

Finally, suppose that $N \lhd G$ and $\vartheta \in \text{Aut}\,G$. Then $|\vartheta(N)| = |N|$ so $|\vartheta(N) \cap N| = \text{h.c.f.}(|\vartheta(N)|, |N|) = |N|$ and hence $\vartheta(N) = N$ as required.

**2.23**   (a) $200 = 5^2 \cdot 2^3$. Hence $G$ contains $k$ Sylow 5–subgroups of order 25 where $k = 1 + 5x$ and $k$ divides 200. Since $(k, 5) = 1$ we have that $k$ divides 8 and hence $x = 0$. Thus $G$ has a unique Sylow 5–subgroup which is therefore normal.

(b) $40 = 5 \cdot 2^3$. Hence $G$ contains $k$ Sylow 5–subgroups where $k = 1 + 5x$ and $k$ divides 40. Here again we have $x = 0$ and so $G$ contains a unique Sylow 5–subgroup which is therefore normal. Hence $G$ is not simple.

(c) $56 = 7 \cdot 2^3$. There are $1 + 7k$ Sylow 7–subgroups where $1 + 7k$ divides 56. If the group is simple then there must be eight Sylow 7–subgroups with 49 distinct elements. Also there must be seven Sylow 2–subgroups and the group now has more than 56 elements.

(d) $35 = 7 \cdot 5$. The number of Sylow 5–subgroups is congruent to 1 mod 5 and divides 35. Hence there is only one Sylow 5–subgroup which is therefore normal. By the same argument, there is only one Sylow 7–subgroup which is therefore normal. Let the Sylow 5–subgroup be $H$ and the Sylow 7–subgroup be $K$. We show as follows that $G \simeq H \times K$.

(i) $H, K \triangleleft G$ has already been seen.

(ii) $|HK| = \dfrac{|H|\,|K|}{|H \cap K|} = \dfrac{5 \cdot 7}{1} = 35$ so $HK = G$.

(iii) $H \cap K = \{1\}$ since h.c.f.$(|H|, |K|) = 1$.

It follows that $G \simeq H \times K \simeq C_5 \times C_7 \simeq C_{35}$ as required.

2.24     (a) If $|G| = 85 = 5 \cdot 17$ then the number of Sylow 5–subgroups is congruent to 1 modulo 5 and so is one of

$$1, \ 6, \ 11, \ 16, \ 21, \ 26, \ \text{etc.}$$

But the number of Sylow 5–subgroups divides 85. Hence $G$ has only one Sylow 5–subgroup, $H$ say, which is then normal. Similarly, there is a unique Sylow 17–subgroup, $K$ say, which is also normal. Now

(i) $H, K \triangleleft G$;

(ii) $G = HK$ since $|HK| = \dfrac{|H|\,|K|}{|H \cap K|} = \dfrac{5 \cdot 17}{1} = 85$;

(iii) $H \cap K = \{1\}$ since h.c.f.$(|H|, |K|) = 1$.

Thus $G \simeq H \times K \simeq C_5 \times C_{17} \simeq C_{85}$.

(b) Let $G$ be a group of order $p^2 q$. Suppose that $G$ is simple. Let $G$ have $n_p$ Sylow $p$–subgroups and $n_q$ Sylow $q$–subgroups. Then $n_p > 1$ and $n_q > 1$. Since $n_p$ divides $q$ we must have $n_p = q$. Also $n_p \equiv 1 \bmod p$ so $q > p$. Again $n_q$ divides $p^2$ so $n_q$ is either $p$ or $p^2$.

There must be $n_q(q-1)$ distinct elements of order $q$. Hence if $n_q = p^2$ there are $p^2 q - p^2(q-1) = p^2$ elements that are not of order $q$. But since the Sylow $p$–subgroup of $G$ has order $p^2$ we have $n_p = 1$, a contradiction. Thus we must have $n_q = p$. But $n_q \equiv 1 \bmod q$ and so $p > q$ which is also a contradiction. We conclude therefore that $G$ cannot be simple.

**2.25** By Sylow's theorem, the number $n$ of distinct Sylow $p$–subgroups of $G$ is a divisor of $q$, and $n \equiv 1 \bmod p$. Since $q$ is prime, we have $n = 1$ or $n = q$. Since $q \not\equiv 1 \bmod p$ it follows that $n \neq 1$. Thus $G$ has a unique Sylow $p$–subgroup $P$, say, and $P \lhd G$.

Consider $S_3$. We have $|S_3| = 2 \cdot 3$ and $3 \equiv 1 \bmod 2$. In this case the result fails since $S_3$ has no normal 2–subgroup.

To show that $G$ is not simple when $|G| = pq$ we can assume that $p > q$, so that $q - 1$ is not divisible by $p$. Then $G$ has a normal Sylow $p$–subgroup so cannot be simple.

Now suppose that $|G| = pq$ where $p \not\equiv 1 \bmod q$ and $q \not\equiv 1 \bmod p$. Then $G$ has a normal Sylow $p$–subgroup $P$ and a normal Sylow $q$–subgroup $Q$. Since $P$ and $Q$ have prime orders they are cyclic. Let $P = \langle x \rangle$ and $Q = \langle y \rangle$. Now $P \cap Q = \{1\}$ so $xy = yx$. Hence $xy$ has order $pq$ and $G = \langle xy \rangle$ is cyclic.

**2.26** The answer is no, and the symmetric group $S_4$ provides an illustration. First we must show that $S_4$ contains a subgroup of order $n$ for every divisor $n$ of 24. The cases $n = 1$ and $n = 24$ are obvious. As for subgroups of order 2, 3, 4, 6, 8, 12 we have

(1) $|\langle(12)\rangle| = 2$;
(2) $|\langle(123)\rangle| = 3$;
(3) $|\langle(1234)\rangle| = 4$;
(4) $S_3 \subset S_4$ and $|S_3| = 6$;
(5) $24 = 3 \cdot 8$ so a Sylow 2–subgroup of $S_4$ has order 8;
(6) $A_4$ has order 12.

Consider now the subgroup $A_4$. We claim that this does not have a subgroup of order 6. To see this, suppose that $H$ were such a subgroup. Then $H$ cannot be abelian (since $S_4$ has no element of order 6) and so $H \simeq S_3$. But every subgroup $S_3$ in $S_4$ fixes a point and contains odd permutations. Hence no such subgroup $H$ can exist.

**2.27** Let $y = g^{-1}xg$ for some $g \in G$. Then $y \in P$ and $y \in g^{-1}Pg$. Hence, since $y \in Z(P)$ and $y \in Z(g^{-1}Pg)$, we have that $P$ and $g^{-1}Pg$ both centralise $y$. Thus
$$\langle P, g^{-1}Pg \rangle \leq N_G(y).$$

But $P$ and $g^{-1}Pg$ are Sylow $p$–subgroups of $G$ and so are Sylow $p$–subgroups of $N_G(y)$. Therefore $P$ and $g^{-1}Pg$ are conjugate in $N_G(y)$. Thus $P = c^{-1}g^{-1}Pgc$ for some $c \in N_G(y)$. Now $gc \in N(P)$ and
$$(gc)^{-1}xgc = c^{-1}g^{-1}xgc = c^{-1}yc = y.$$

Hence $x$ and $y$ are conjugate in $N(P)$.

# Solutions to Chapter 2

**2.28**  Let $\Omega = \{aH \mid a \in G\}$ and for every $g \in G$ define a permutation $\bar{g}$ on $\Omega$ by

$$\bar{g} : aH \mapsto gaH.$$

Then $\varphi : G \to S_\Omega$ defined by $\varphi(g) = \bar{g}$ is readily seen to be a group morphism. Now $g \in \operatorname{Ker} \varphi$ if and only if $aH = gaH$ for all $a \in G$, which is the case if and only if $g \in aHa^{-1}$ for all $a \in G$. Thus we see that $\operatorname{Ker} \varphi = \bigcap_{a \in G} a^{-1}Ha$ which is easily seen to be the largest normal subgroup of $G$ contained in $H$. By the first isomorphism theorem,

$$G/\operatorname{Ker} \varphi \simeq \operatorname{Im} \varphi \leq S_n.$$

Now suppose that $G$ is simple with $|G| = 60$. If $H$ is a subgroup with $|H| = 15$ then $H$ has index 4. But $K = \{1\}$ since $G$ is simple, and $G/K = G$ which (by the above) must be isomorphic to a subgroup of $S_4$. However, 60 does not divide 24, and so we have a contradiction.

In a similar way we can show that $G$ has no subgroups of order 20 or 30.

**2.29**  The number $n$ of Sylow 7–subgroups is such that $n \equiv 1 \bmod 7$ and $n$ divides 168. Now $n \neq 1$ since otherwise $G$ has a unique Sylow 7–subgroup which is therefore normal. The only other divisor of 168 that is congruent to 1 modulo 7 is 8. Hence $G$ has eight Sylow 7–subgroups.

Since $\mathcal{N}_G(P)$ must have index 8, we have $|\mathcal{N}_G(P)| = 21$.

Suppose now that $H \leq G$ with $|H| = 14$. We derive a contradiction as follows. Consider the number $m$ of Sylow 7–subgroups of $H$. We have that $m \equiv 1 \bmod 7$ and $m$ divides 14. Thus $m = 1$ and $H$ has a normal Sylow 7–subgroup $K$. However, $|K| = 7$ so $K$ must be a Sylow 7–subgroup of $G$. Now since $K$ is normal in $H$ we must have $H \leq \mathcal{N}_G(K)$. This shows that $|H| = 14$ divides $|\mathcal{N}_G(K)| = 21$ which is the required contradiction.

# Solutions to Chapter 3

3.1     (a) Given $[s,t] \in [S,T]$ we have $[s,t]^{-1} = [t,s] \in [T,S]$. Hence $[S,T] \subseteq [T,S]$ and similarly for the reverse inclusion.

    (b) We have that

$$h^{-1}k^{-1}hk = h^{-1}h' \in H \quad \text{since } H \triangleleft G$$

and

$$h^{-1}k^{-1}hk = k'k \in K \quad \text{since } K \triangleleft G.$$

Hence $[H,K] \leq H \cap K$. In the case where $H \cap K = \{1\}$ we have $[H,K] = \{1\}$ and the elements of $H$ commute with those of $K$.

    (c) The first part follows from $[xy,z] = y^{-1}x^{-1}z^{-1}xyz$ and

$$y^{-1}[x,z]y[y,z] = y^{-1}x^{-1}z^{-1}xzyy^{-1}z^{-1}yz.$$

Clearly $[H,K] \leq [HL,K]$ and $[L,K] \leq [HL,K]$, so

$$[H,K][L,K] \leq [HL,K].$$

But $[hl,k] = l^{-1}[h,k]l[l,k] \in [H,K][L,K]$ since $[H,K] \triangleleft G$. Hence the required equality follows.

    (d) This follows immediately on expanding the commutators.

3.2     Let $Q_8 = \langle a,b \mid a^2 = b^2 = (ab)^2 \rangle$ and $C_2 = \langle c \mid c^2 = 1 \rangle$. Then $Z(G) = \langle a^2, c \rangle$ and the upper central series is

$$\{1\} < \langle a^2, c \rangle < G.$$

The derived group of $G$ is

$$G' = \langle a^2 \rangle$$

and the lower central series is

$$G > \langle a^2 \rangle > \{1\}.$$

Thus $G$ is nilpotent of class 2 and the upper and lower central series do not coincide.

$Q_8$ has derived group $\langle a^2 \rangle$ and $Z(Q_8) = \langle a^2 \rangle$. Thus the upper and lower central series of $Q_8$ are both equal to

$$Q_8 > \langle a^2 \rangle > \{1\}.$$

3.3   Suppose that $G$ is generated by its subnormal abelian subgroups. Then

$$G = \langle G_\lambda \mid G_\lambda \text{ subnormal abelian} \rangle.$$

Now since $G_\lambda$ is subnormal in $G$ we have a series

$$G_\lambda = R_0 \leq R_1 \leq \cdots \leq R_n = G$$

with $R_i$ normal in $R_{i+1}$ for $0 \leq i \leq n-1$. But if $H$ is a quotient group of $G$ then $H \simeq G/K$ and so

$$\frac{G_\lambda K}{K} = \frac{G_\lambda R_0}{K} \leq \frac{G_\lambda R_1}{K} \leq \cdots \leq \frac{G_\lambda R_n}{K} = \frac{G}{K}$$

is a series of subgroups with $\dfrac{G_\lambda R_i}{K}$ normal in $\dfrac{G_\lambda R_{i+1}}{K}$ for $0 \leq i \leq n-1$.
Now

$$\frac{G_\lambda K}{K} \simeq \frac{G_\lambda}{G_\lambda \cap K}$$

which is abelian since $G_\lambda$ is abelian. Hence

$$\frac{G}{K} = \left\langle \frac{G_\lambda K}{K} \;\middle|\; \frac{G_\lambda K}{K} \text{ subnormal abelian in } \frac{G}{K} \right\rangle.$$

Therefore $H \simeq G/K$ is generated by its subnormal abelian subgroups.

Suppose now that $G$ is a nilpotent group and that $H \leq G$. Let

$$G = G_0 \geq G_1 \geq \cdots \geq G_r = \{1\}$$

be a central series for $G$. Consider the series

$$H = HG_r \leq HG_{r-1} \leq \cdots \leq HG_1 \leq HG_0 = G.$$

Now $HG_i \lhd HG_{i-1}$, since for $h, h' \in H, g_i \in G_i, g_{i-1} \in G_{i-1}$ we have

$$(h'g_{i-1})^{-1} h g_i h' g_{i-1} = g_{i-1}^{-1} h'^{-1} h g_i h' g_{i-1}.$$

But, for any $x \in G$, $x g_{i-1} = g_{i-1} x g_i'$ for some $g_i' \in G_i$ so it follows that $g_{i-1}^{-1} h'^{-1} h g_i h' g_{i-1} \in HG_i$ as required. Since every group is generated by abelian subgroups (the cyclic subgroups generated by its elements), a nilpotent group is then generated by its subnormal abelian subgroups.

3.4    Apply the fundamental isomorphism

$$\frac{H}{H \cap K} \simeq \frac{HK}{K}$$

to the case where $G = A$, $H = A \cap C$ and $K = B$; then since $B \triangleleft A$ we obtain

$$\frac{A \cap C}{B \cap C} \simeq \frac{(A \cap C)B}{B}.$$

Now apply the isomorphism to the case where $G = AC, H = A$ and $K = BC$ to obtain

$$\frac{A}{A \cap BC} \simeq \frac{A(BC)}{BC}.$$

But $A \cap BC = B(A \cap C)$ and $A(BC) = AC$, and the result follows.

Suppose that $G$ is a soluble group with a series

$$\{1\} = G_0 \leq G_1 \leq \cdots \leq G_n = G$$

where each $G_i/G_{i-1}$ is abelian. Let $H \leq G$ and consider the series

$$\{1\} = G_0 \cap H \leq G_1 \cap H \leq \cdots \leq G_n \cap H = H.$$

We have

$$\frac{G_i \cap H}{G_{i-1} \cap H} \simeq \frac{G_{i-1}(G_i \cap H)}{G_{i-1}} \leq \frac{G_i}{G_{i-1}}$$

and so $\dfrac{G_i \cap H}{G_{i-1} \cap H}$ is abelian (being a subgroup of an abelian group).

Also, if $K \triangleleft G$ then

$$\frac{K}{K} = \frac{G_0 K}{K} \leq \frac{G_1 K}{K} \leq \cdots \leq \frac{G_n K}{K} = \frac{G}{K}.$$

Now we have

$$\frac{G_i K/K}{G_{i-1}K/K} \simeq \frac{G_i K}{G_{i-1}K} \simeq \frac{G_i}{G_{i-1}(G_i \cap K)}$$

which is a quotient group of $\dfrac{G_i}{G_{i-1}}$. Hence $\dfrac{G_i K/K}{G_{i-1}K/K}$ is abelian (being a quotient group of an abelian group).

If $H \triangleleft K$ and both $H$ and $K/H$ are soluble then we have

$$\frac{K}{H} = \frac{K_0}{H} \geq \frac{K_1}{H} \geq \cdots \geq \frac{K_r}{H} = \frac{H}{H}$$

and $H = H_0 \geq H_1 \geq \cdots \geq H_s = \{1\}$, whence we have that

$$K = K_0 \geq K_1 \geq \cdots \geq K_r = H = H_0 \geq H_1 \geq \cdots \geq H_s = \{1\}$$

is a series for $K$ with abelian factors.

For the last part, suppose that $G/A$ and $G/B$ are soluble. Then

$$\frac{A}{A \cap B} \simeq \frac{AB}{B} \leq \frac{G}{B}$$

and so $A/(A \cap B)$ is soluble. Then $G/A$ soluble and $A/(A \cap B)$ soluble gives $G/(A \cap B)$ soluble.

**3.5**   (a) True. We have $HK/K \simeq H/(H \cap K)$ which is soluble, being a quotient group of the soluble group $H$. But then $HK/K$ is soluble and $K$ is soluble, so $HK$ is soluble.

(b) False. For example, consider

$$G = Q_8 = \langle\, a, b \mid a^2 = b^2 = (ab)^2 \,\rangle.$$

$H = \langle\, a \,\rangle$ and $K = \langle\, b \,\rangle$ are normal cyclic subgroups but $HK = Q_8$ which is not abelian.

(c) True, by the same argument as in (a).

**3.6**   Let $H$ be a proper subgroup of $G$. If $Z(G) \nsubseteq H$ then $Z(G)H$ normalises $H$ and the result follows. Suppose then that $Z(G) \subseteq H$. Suppose, by way of induction, that the result holds for groups of order less than $|G|$. Since $Z(G) \neq \{1\}$ we can apply the induction hypothesis to $H/Z(G)$ as a subgroup of $G/Z(G)$. This shows that $H/Z(G)$ is properly contained in its normaliser, $K/Z(G)$ say. Then $H$ is normal in $K$ and properly contained in $K$ as required.

Let $P$ be a Sylow subgroup of $G$. Using the result of question 2.21, we have that $\mathcal{N}(P)$ is equal to its own normaliser in $G$ and so, by the first part of the question, $\mathcal{N}(P)$ cannot be a proper subgroup of $G$. Hence $\mathcal{N}(P) = G$ and so $P$ is normal in $G$.

**3.7**   If $|Z(G)| > 2$ then $|G/Z(G)| \leq 4$ and so $G/Z(G)$ is abelian. This contradicts the class of $G$ being 3. Hence we have that $|Z(G)| = 2$ and $|G/Z(G)| = 8$. Now $G/Z(G)$ must contain an element $xZ(G)$ of order 4, otherwise $G/Z(G)$ would be abelian. Let $H = \langle\, x, Z(G) \,\rangle$ so that $H$ is an abelian subgroup of $G$ with $|H| = 8$. We show first that $H$ is cyclic.

Suppose that $H$ is not cyclic, so that $H = \langle\, x \,\rangle \times Z(G)$. Consider the subgroup $\langle\, x^2 \,\rangle$. There must be some $g \in G$ such that $[x^2, g] \neq 1$, for otherwise $x^2 \in Z(G)$. Now consider the non-trivial element $x^{-2}g^{-1}x^2g$. Since $H$ is normal in $G$, we have that $g^{-1}xg \in H$ and $g^{-1}x^2g$ must be a non-trivial square of an element of $H$. This gives $g^{-1}x^2g = x^2$ and so $[x^2, g] = 1$, a contradiction.

Suppose now that $H$ and $K$ are cyclic subgroups of order 8 with $H \neq K$. Then $G = HK$ and so $H \cap K \leq Z(G)$. However, this gives $|H \cap K| \leq 2$, which contradicts

$$|HK||H \cap K| = |H||K|.$$

Thus we conclude that $H = K$.

An example of such a group is the dihedral group $D_{16}$.

*3.8*    If $G$ is a finite nilpotent group then it has a lower central series which satisfies the conditions required for $G$ to be residually nilpotent. Conversely, if $G$ is finite and residually nilpotent then $H_i$ cannot properly contain $H_{i+1}$ except for finitely many $i$, so $H_i = H_{i+1}$ for all $i \geq N$. But now, since

$$\bigcap_{i=1}^{\infty} H_i = \bigcap_{i=1}^{N} H_i = H_N,$$

we have $H_N = \{1\}$ and so $G$ is nilpotent.

The group

$$D_{\infty} = \langle\, a, b \mid b^2 = 1, bab = a^{-1} \,\rangle$$

is residually nilpotent. For, taking $H_i = \langle\, a^{2^i} \,\rangle$ we have that $[H_i, G] \leq H_{i+1}$ and $\bigcap_{i=1}^{\infty} H_i = \{1\}$. However, $D_{\infty}$ is not nilpotent. To see this, take $K = \langle\, a^3 \,\rangle$ and observe that

$$D_{\infty}/K \simeq \langle\, a, b \mid a^3 = b^2 = 1, bab = a^{-1} \,\rangle \simeq S_3$$

which is not nilpotent.

If $H$ is a subgroup of a residually nilpotent group $G$ then intersecting the series of $G$ with $H$ shows that $H$ is residually nilpotent (the argument generalises the usual proof that a subgroup of a nilpotent group is nilpotent).

Consider $D_{\infty}$ again. We have seen above that $S_3$ is a quotient group. But since $S_3$ is finite and not nilpotent, it fails to be residually nilpotent (by the first part of the question). Thus we see that a quotient group of a residually nilpotent group need not be residually nilpotent. (In fact, every group is a quotient group of a residually nilpotent group.)

# Solutions to Chapter 3

**3.9** It is readily seen by expanding the commutators that
$$[xy, z] = y^{-1}[x, z]y[y, z].$$
Now $[G, A]$ is generated by $[g, a]$ where $g \in G$ and $a \in A$. It is therefore sufficient to check that $x^{-1}[g, a]x \in [G, A]$ for all $g, x \in G$ and $a \in A$. But
$$x^{-1}[g, a]x = [gx, a]\,[x, a]^{-1} \in [G, A]$$
and so $[G, A] \triangleleft G$.

Suppose that $A = [A, G]$ and that $G$ is nilpotent of class $n$ say. Then we have
$$A = [A, \underbrace{G, G, \ldots, G}_{n}] = \{1\}.$$
Thus if $A \neq \{1\}$ then $G$ cannot be nilpotent.

Let $A$ be a minimal normal subgroup of a nilpotent group $G$. Then $[G, A] \leq A$ and so, since $[G, A] \triangleleft G$, we must have either $[G, A] = A$ or $[G, A] = \{1\}$. The former is impossible since $G$ is nilpotent. Hence $[G, A] = \{1\}$ and $A$ is in the centre of $G$.

**3.10** Let $G$ be a finite $p$-group. First we show that the centre of $G$ is nontrivial. Note that a conjugacy class has one element if and only if the elements of the class are central. Now $G$ is the union of its conjugacy classes, and $\{1\}$ is a conjugacy class. Any conjugacy class containing more than one element has $k$ elements where $p | k$. Hence $G$ has more than one conjugacy class containing one element and so $Z(G) \neq \{1\}$.

Let $Z_2(G)/Z(G)$ be the centre of $G/Z(G)$. Continuing in this way, we obtain a series
$$\{1\} \leq Z(G) \leq Z_2(G) \leq \cdots \leq G$$
and, since $G$ is finite, $Z_n(G) = G$ for some $n$. Thus $G$ is nilpotent.

Suppose that $G = H \times K$ where $|H| = p^2$ and $|K| = p^3$. Now we have that $Z(G) = Z(H) \times Z(K)$ and $Z(H) = H$ since a group of order $p^2$ is abelian. Consequently, $|Z(G)| = p^2|Z(K)|$. It follows that $|Z(K)|$ is $p$, or $p^2$, or $p^3$; for $|Z(K)| \neq 1$ since $K$ is a $p$-group. But if $|Z(K)| = p^2$ then $|K/Z(K)| = p$ and so $K/Z(K)$ is cyclic, say $K/Z(K) = \langle\, aZ(K) \,\rangle$. Then $k \in K$ gives $k = a^i z$ for some $z \in Z(K)$. Then any two elements of $K$ commute, whence it follows that $|Z(K)| = p^3$ which contradicts the hypothesis that $|Z(K)| = p^2$. We now note that $|Z(K)| \neq p^3$ since otherwise $G$ is abelian. Thus we have that $|Z(K)|$ must be $p$, whence $|Z(G)| = p^3$ and $|G/Z(G)| = p^2$. Using again the fact that a group of order $p^2$ is abelian, we see that
$$G > Z(G) > \{1\}$$
is the upper central series of $G$, and hence that $G$ is nilpotent of class 2.

*3.11*    Consider the matrices

$$x = \begin{bmatrix} 1 & a & b \\ 0 & 1 & c \\ 0 & 0 & 1 \end{bmatrix}, \quad y = \begin{bmatrix} 1 & d & e \\ 0 & 1 & f \\ 0 & 0 & 1 \end{bmatrix}$$

where $a, b, c, d, e, f \in \mathbb{Z}$. Then $[x, y]$ is given by

$$\begin{bmatrix} 1 & -a & ac-b \\ 0 & 1 & -c \\ 0 & 0 & 1 \end{bmatrix} \begin{bmatrix} 1 & -d & df-e \\ 0 & 1 & -f \\ 0 & 0 & 1 \end{bmatrix} \begin{bmatrix} 1 & a & b \\ 0 & 1 & c \\ 0 & 0 & 1 \end{bmatrix} \begin{bmatrix} 1 & d & e \\ 0 & 1 & f \\ 0 & 0 & 1 \end{bmatrix}$$

which reduces to

$$\begin{bmatrix} 1 & 0 & af-dc \\ 0 & 1 & 0 \\ 0 & 0 & 1 \end{bmatrix}.$$

Now if $x$ is central in $G$ then $[x, y] = I_3$ for all $y \in G$, which gives $af - dc = 0$ for all $d, f \in \mathbb{Z}$, whence $a = c = 0$. Thus we see that the centre of $G$ is

$$Z = \left\{ \begin{bmatrix} 1 & 0 & b \\ 0 & 1 & 0 \\ 0 & 0 & 1 \end{bmatrix} \mid b \in \mathbb{Z} \right\}.$$

The derived group of $G$ is $\langle [x, y] \mid x, y \in G \rangle$. From the above calculation of $[x, y]$ it is easy to see that the derived group of $G$ is $Z$. To see that $G$ is nilpotent, we show that $[Z, G] = \{I_3\}$. Again, this follows from the above calculation.

The upper central series is

$$\{I_3\} < Z < G.$$

For, we have shown that $Z$ is the centre of $G$, and $G/Z$ is abelian since $Z$ is the derived group of $G$. This is also the lower central series of $G$ as our calculations have shown. Hence the upper and lower central series of $G$ coincide.

For the given matrices we have

$$t_{12}^n = \begin{bmatrix} 1 & n & 0 \\ 0 & 1 & 0 \\ 0 & 0 & 1 \end{bmatrix}, \quad t_{13}^n = \begin{bmatrix} 1 & 0 & n \\ 0 & 1 & 0 \\ 0 & 0 & 1 \end{bmatrix}, \quad t_{23}^n = \begin{bmatrix} 1 & 0 & 0 \\ 0 & 1 & n \\ 0 & 0 & 1 \end{bmatrix}.$$

Now since

$$
\begin{bmatrix} 1 & a & b \\ 0 & 1 & c \\ 0 & 0 & 1 \end{bmatrix} = \begin{bmatrix} 1 & a & 0 \\ 0 & 1 & 0 \\ 0 & 0 & 1 \end{bmatrix} \begin{bmatrix} 1 & 0 & 0 \\ 0 & 1 & c \\ 0 & 0 & 1 \end{bmatrix} \begin{bmatrix} 1 & 0 & b-ac \\ 0 & 1 & 0 \\ 0 & 0 & 1 \end{bmatrix}
$$

it follows that we have $G = \langle t_{12}, t_{13}, t_{23} \rangle$.

A subnormal series for $\langle t_{12} \rangle$ is

$$\langle t_{12} \rangle \lhd \langle t_{12}, t_{13} \rangle \lhd G.$$

A subnormal series for $\langle t_{23} \rangle$ is

$$\langle t_{23} \rangle \lhd \langle t_{23}, t_{13} \rangle \lhd G.$$

A subnormal series for $\langle t_{13} \rangle$ is

$$\langle t_{13} \rangle \lhd G.$$

**3.12** Suppose that $N \lhd G$ and $A \leq N, B \leq N$. Then for all $x \in X, y \in Y, z \in Z$ we have

$$[x, y^{-1}, z]^y \in N \quad \text{and} \quad [y, z^{-1}, x]^z \in N.$$

Hence $[z, x^{-1}, y]^x \in N$ and so $[z, x^{-1}, y] \in N$, whence $[z, x^{-1}]$ commutes with $y$ modulo $N$, whence $[Z, X]$ commutes with $Y$ element-wise modulo $N$, and consequently $C \leq N$.

For the next part, use induction. The result is clearly true if $n = 1$. Assume then that $[H_i, \Gamma_{n-1}(K)] \leq H_{i+n-1}$. Let

$$X = \Gamma_{n-1}(K), \quad Y = K, \quad Z = H_i, \quad N = H_{i+n}.$$

Then we have

$$A = [\Gamma_{n-1}(K), K, H_i] = [\Gamma_n(K), H_i]$$
$$B = [K, H_i, \Gamma_{n-1}(K)] \leq [H_{i+1}, \Gamma_{n-1}(K)] \leq H_{i+n} = N$$
$$C = [H_i, \Gamma_{n-1}(K), K] \leq [H_{i+n-1}, K] \leq H_{i+n} = N$$

and hence $A \leq N$ as required.

For the last part, take $G = H = K$ and the series to be the lower central series. Then

(a) and (b) follow immediately;

(c) follows from (b) with $m = n$;

(d) is proved by induction. We have $G^{(0)} = \Gamma_1$. Suppose that $G^{(r-1)} \leq \Gamma_{2^{r-1}}$; then, by (a),

$$G^{(r)} = [G^{(r-1)}, G^{(r-1)}] \leq [\Gamma_{2^{r-1}}, \Gamma_{2^{r-1}}] \leq \Gamma_{2^r}.$$

(e) $[Z_2, \Gamma_2] = \{1\}$ and so $[Z_2, G] = \{1\}$ gives $Z_2 \leq Z_1$ whence $Z_2 = Z_1$.

*3.13*   Since $x \in Z_2(G)$ we have $[x, g] \in Z(G)$ so $N \leq Z(G)$. Now in this case we have

$$[x, g_1][x, g_2] = x^{-1} g_1^{-1} x g_1 x^{-1} g_2^{-1} x g_2$$
$$= x^{-1} g_2^{-1} x . x^{-1} g_1^{-1} x g_1 . g_2$$
$$= [x, g_1 g_2]$$

and hence

$$N = \{[x, g] \mid g \in G\}.$$

Now

$$[x, g_1] = [x, g_2] \iff g_1^{-1} x g_1 = g_2^{-1} x g_2$$
$$\iff \mathcal{N}(x) g_1 = \mathcal{N}(x) g_2.$$

This gives $|N| = |G : \mathcal{N}(x)|$. However, $Z(G) \leq \mathcal{N}(x)$ and $x \notin Z(G), x \in \mathcal{N}(x)$ and so $Z(G)$ is properly contained in $\mathcal{N}(x)$, whence $|N| < p^n$.

If $n = 1$ then $G$ is abelian and $G' = \{1\}$ so the result holds. Suppose, by way of induction, that the result holds for all groups with factor by the centre of order $p^k$ for $k < n$. Consider $Z(G/N)$. Certainly $Z(G)/N \leq Z(G/N)$. But $xN \in Z(G/N)$ since $[x, G] \subseteq N$ so $Z(G/N)$ properly contains $Z(G)/N$. Hence

$$|G/N : Z(G/N)| = p^k$$

for some $k < n$, and so by the inductive hypothesis

$$|G'/N| \leq p^{\frac{1}{2}(n-1)(n-2)}.$$

But $|N| \leq p^{n-1}$ and so $|G'| \leq p^{\frac{1}{2}n(n-1)}$ as required.

*3.14*   Suppose that $H$ is a subgroup of $G$ with $G = \Phi H$. Then if $H \neq G$ we have that $H$ is contained in a maximal subgroup $M$ of $G$, whence $G = \Phi H \leq \Phi M$ and consequently $G = \Phi M$. But $\Phi \leq M$ since $\Phi$ is the intersection of all the maximal subgroups of $G$. Hence $G = \Phi M \leq M$ which is a contradiction. Thus $H = G$ as required.

Since $T \leq \Phi$ we have $T^g \leq \Phi^g$. First we show that $\Phi$ is normal in $G$. If $M$ is a maximal subgroup of $G$ then $g^{-1} M g$ is also maximal, since otherwise $g^{-1} M g < K < G$ for some subgroup $K$ and then $M < gKg^{-1} < G$, a contradiction. Suppose now that $x \in \Phi$. If $g^{-1} x g \notin \Phi$ then $g^{-1} x g \notin M$ for some maximal subgroup $M$, and then $x \notin gMg^{-1}$ which is a contradiction. Therefore $T^g \leq \Phi^g = \Phi$ and so $T$ and $T^g$ are Sylow $p$–subgroups of $\Phi$. Thus $T^g = T^h$ for some $h \in \Phi$, by Sylow's theorem. It now follows that $T^{gh^{-1}} = T$ and so $gh^{-1} \in \mathcal{N}_G(T)$ whence

$g \in \mathcal{N}_G(T)\Phi$. Hence we see that $\mathcal{N}_G(T)\Phi = G$ and, by the first part of the question, that $\mathcal{N}_G(T) = G$. Thus $T$ is normal in $G$ and so is normal in $\Phi$.

Thus every Sylow $p$–subgroup of $\Phi$ is normal.

**3.15**  Suppose that $G$ satisfies the maximum condition for subgroups and let $H$ be a subgroup of $G$. Choose $x_1 \in H$ and let $H_1 = \langle x_1 \rangle$. If $H_1 < H$ choose $x_2 \in H \setminus H_1$ and let $H_2 = \langle x_1, x_2 \rangle$. Continuing in this way, we obtain a chain of subgroups

$$H_1 < H_2 < \cdots < H_n < H_{n+1} < \cdots$$

in which, by the maximum condition, $H_r = H$ for some $r$. Hence

$$H = H_r = \langle x_1, x_2, \ldots, x_r \rangle$$

is finitely generated.

Conversely, if $G$ fails to satisfy the maximum condition then $G$ contains an infinite chain of distinct subgroups

$$H_1 < H_2 < \cdots < H_n < H_{n+1} < \cdots .$$

Let $H = \bigcup_{i \geq 1} H_i$ and suppose that $H$ is finitely generated. If, say, $H = \langle x_1, \ldots, x_r \rangle$ then we have

$$x_1 \in H_{s_1}, \quad x_2 \in H_{s_2}, \ldots, x_r \in H_{s_r}$$

and consequently each $x_i \in H_s$ where $s = \max_{1 \leq i \leq r} s_i$. It follows that $H = H_s$, a contradiction. Hence $H$ cannot be finitely generated.

Let $G$ be a soluble group that satisfies the maximum condition for subgroups. Let

$$G = H_0 \geq H_1 \geq \cdots \geq H_r = \{1\}$$

be a series for $G$ with each quotient $H_{i-1}/H_i$ abelian. Now $H_{i-1}$, being a subgroup of $G$, is finitely generated and so therefore is $H_{i-1}/H_i$. Thus we may insert between $H_i$ and $H_{i-1}$ a finite number of subgroups to obtain a series in which quotients of consecutive members are cyclic. Thus we have that $G$ is polycyclic.

Conversely, if $G$ is polycyclic let

$$G = H_0 \geq H_1 \geq \cdots \geq H_r = \{1\}$$

be a series with each quotient $H_{i-1}/H_i$ cyclic. Then if $K \leq G$ we have, writing $K_i = H \cap H_i$, the series

$$K = K_0 \geq K_1 \geq \cdots \geq K_r = \{1\}$$

in which consecutive quotients are cyclic. Let $K_i a_{i-1}$ be a generator of $K_{i-1}/K_i$. Then it follows that

$$K = \langle a_0, a_1, \ldots, a_{r-1} \rangle$$

and so $K$ is finitely generated as required.

**3.16**   Rewriting the given relation in the form

(1) $\qquad\qquad y^{-1}[x,z]y = [xy,z][y,z]^{-1}$

we see that, for all $x, y \in A$ and all $z \in B$,

$$y^{-1}[x,z]y \in [A,B]$$

and so $A$ normalises $[A,B]$. Similarly, $B$ normalises $[A,B]$ and so $[A,B]$ is normal in $G$.

Replacing $z$ by $zt$ in (1) gives

$$[xy,zt] = [x,zt]^y[y,zt].$$

However, $[x,zt] = [zt,x]^{-1}$ and $[y,zt] = [zt,y]^{-1}$ so we can use (1) again to express $[xy,zt]$ as a product of conjugates of commutators of the form $[a,b]$ where $a,b \in \{x,y,z,t\}$. Hence if $x,z \in A$ and $y,t \in B$ then using the fact that $[x,z] = 1 = [y,t]$ and the fact that $[A,B]$ is normal in $G$ we see that $[xy,zt] \in [A,B]$. Thus $G' \subseteq [A,B]$. However, $[A,B] \subseteq [AB,AB] = G'$ and so $G' = [A,B]$.

Since $AB = BA$ we have

$$b_1^{a_2} = a_3 b_3 \quad \text{and} \quad a_1^{b_2} = b_4 a_4$$

for some $a_3, a_4 \in A$ and $b_3, b_4 \in B$. Now

$$
\begin{aligned}
[a_1,b_1]^{a_2 b_2} = [a_1^{a_2}, b_1^{a_2}]^{b_2} &= [a_1, a_3 b_3]^{b_2} \\
&= [a_1^{b_2}, b_3] = [b_4 a_4, b_3] \\
&= [a_4, b_3].
\end{aligned}
$$

Similarly we can show that

$$[a_1,b_1]^{b_2 a_2} = [a_4, b_3]$$

and so $[a_1,b_1]^{a_2 b_2} = [a_1,b_1]^{b_2 a_2}$ as required. It now follows that

$$[a_1,b_1]^{[b_2,a_2]} = [a_1,b_1]$$

and so $[A,B]$ is abelian.

The derived series for $G$ is now

$$G \geq [A,B] \geq \{1\}$$

and so $G$ is soluble, of derived length at most 2.

**3.17** Let $M$ be a maximal subgroup of $G$. Then $M$ is subnormal in $G$ (see question 3.3) so we have

$$M \lhd H_1 \lhd \cdots \lhd H_r \lhd G.$$

But $M < H_1 < G$ is impossible, so $M \lhd G$.

$S$ is a Sylow $p$-subgroup and $M$ is a maximal subgroup of $G$ with

$$\mathcal{N}_G(S) \leq M < G.$$

Let $g \in G$. Then $g^{-1}Sg \leq g^{-1}Mg$. But $g^{-1}Mg = M$ since $M$ is normal. Therefore $S$ and $g^{-1}Sg$ are Sylow $p$-subgroups of $M$. Now, using the Sylow theorems, we have that $S$ and $g^{-1}Sg$ are conjugate in $M$. Hence there exists $m \in M$ with $g^{-1}Sg = m^{-1}Sm$. This shows that $mg^{-1}Sgm^{-1} = S$ and so $(gm^{-1})^{-1}Sgm^{-1} = S$ whence $gm^{-1} \in \mathcal{N}_G(S)$. It now follows that $gm^{-1} \in M$ whence $g \in M$ since $m \in M$. This is clearly impossible since we now have the contradiction $M = G$. We deduce, therefore, that $\mathcal{N}_G(S) = G$ and so $S$ is normal in $G$.

Suppose now that $\overline{S}$ is a Sylow $p$-subgroup of $G$. Then $\overline{S} = g^{-1}Sg$ for some $g \in G$, by Sylow's theorem. Hence $\overline{S} = g^{-1}Sg = S$ since $S \lhd G$ and so $S$ is unique.

Let $S_1, \ldots, S_r$ be Sylow subgroups, each corresponding to distinct prime divisors of $|G|$. Since $S_i$ is the only Sylow subgroup for its associated prime, we have $S_i \lhd G$. Clearly, if $i \neq j$ then $S_i \cap S_j = \{1\}$ since h.c.f.$(|S_i|, |S_j|) = 1$. Hence

$$G = S_1 \times S_2 \times \cdots \times S_r.$$

**3.18** Let $k \in K$. Then $k \in g^{-1}Mg$ for every $g \in G$. If $x \in G$ we then have

$$(\forall g \in G) \qquad x^{-1}kx \in (gx)^{-1}Mgx$$

and so $x^{-1}kx$ belongs to every conjugate of $M$. Hence $x^{-1}kx \in K$ and so $K \lhd G$.

If $N \lhd G$ and $N \leq M$ then for every $g \in G$ we have

$$N = g^{-1}Ng \leq g^{-1}Mg$$

from which it follows that $N \leq K$.

Since $H/K$ is a minimal normal subgroup of $G/K$ we have $H \lhd G$ and $K \subset H$, so $H \not\leq M$. Therefore $G = MH$ since $M$ is maximal and $G \geq MH > M$.

Now $H \cap M$ is normal in $M$ and so $(H \cap M)/K$ is normal in $M/K$. But $(H \cap M)/K$ is normal in $H/K$ since $H/K$ is a minimal normal subgroup of a soluble group and is therefore abelian. Thus every subgroup of $H/K$ is normal.

Now $(H \cap M)/K$ is normalised by $M/K$ and is also normalised by $H/K$. Hence $H/K \cdot M/K$ normalises $(H \cap M)/K$. But

$$H/K \cdot M/K = HM/K = G/K.$$

Consequently, $(H \cap M)/K$ is normal in $G/K$. It now follows that $H \cap M$ is normal in $G$.

Now $H \cap M \leq M$ and $H \cap M \triangleleft G$ imply that $H \cap M \leq K$. But $K \leq H$ and $K \leq M$, so we have $H \cap M = K$.

Finally, $|MH : M| = |H : H \cap M|$ gives $|G : M| = |H : K|$.

**3.19** Suppose that $G$ is metacyclic and that $N \triangleleft G$ with $N$ and $G/N$ cyclic. Let $H \leq G$. Then $H \cap N \triangleleft H$ and $H \cap N$ is cyclic (since it is a subgroup of the cyclic group $N$). Also,

$$H/(H \cap N) \simeq HN/N \leq G/N$$

so $H/(H \cap N)$ is cyclic (being isomorphic to a subgroup of the cyclic group $G/N$). Hence $H$ is metacyclic.

Suppose now that $K \triangleleft G$. We have $K \triangleleft NK$ and $NK/K \simeq N/(N \cap K)$ which is cyclic (being a quotient group of the cyclic group $N$). Also,

$$\frac{G/K}{NK/K} \simeq \frac{G}{NK} \simeq \frac{G/N}{NK/N}$$

which is cyclic (being a quotient group of the cyclic group $G/N$). Hence $\frac{NK}{K} \triangleleft \frac{G}{K}$ with $\frac{NK}{K}$ cyclic and $\frac{G/K}{NK/K}$ cyclic, so $G/K$ is metacyclic. If

$$G = \langle a, b \mid a^2 = 1, \ b^8 = 1, \ aba = b^7 \rangle$$

then we have $a^{-1}ba = b^{-1}$ so $\langle b \rangle \triangleleft G$ since then $a^{-1}b^i a = b^{-i}$ for all integers $i$. But

$$G/\langle b \rangle = \langle a, b \mid a^2 = 1 = b \rangle$$

and so $G/\langle b \rangle \simeq C_2$. Also, $\langle b \rangle \simeq C_8$ so $G$ is metacyclic.

*3.20*   Consider the following matrices (in each of which the entries not shown are all 0) :

$$A_0 = \begin{bmatrix} 1 & a_{12} & a_{13} & \cdots & a_{1n} \\ & 1 & a_{23} & \cdots & a_{2n} \\ & & 1 & \cdots & a_{3n} \\ & & & \ddots & \vdots \\ & & & & 1 \end{bmatrix}$$

$$A_1 = \begin{bmatrix} 1 & -a_{12} & -a_{13} & \cdots & -a_{1n} \\ & 1 & & & \\ & & 1 & & \\ & & & \ddots & \\ & & & & 1 \end{bmatrix}$$

$$A_2 = \begin{bmatrix} 1 & & & & \\ & 1 & -a_{23} & \cdots & -a_{2n} \\ & & 1 & & \\ & & & \ddots & \\ & & & & 1 \end{bmatrix}$$

$$\vdots$$

$$A_{n-1} = \begin{bmatrix} 1 & & & & \\ & 1 & & & \\ & & \ddots & & \\ & & & & -a_{n-1,n} \\ & & & & 1 \end{bmatrix}.$$

It is readily seen that $A_0 A_1 \cdots A_{n-1} = I_n$. Also,

$$\begin{bmatrix} 1 & -a_{12} & -a_{13} & \cdots & -a_{1n} \\ & 1 & & & \\ & & 1 & & \\ & & & \ddots & \\ & & & & 1 \end{bmatrix} = t_{12}(-a_{12}) t_{13}(-a_{13}) \cdots t_{1n}(-a_{1n}).$$

Expanding the other matrices in a similar way, the description of $T_n(F)$ follows.

A similar argument shows that $H$ is the set of all upper triangular matrices over $F$ of the form

$$\begin{bmatrix} 1 & & & a_{1,i+1} & a_{1,i+2} & \cdots & a_{1n} \\ & 1 & & & a_{2,i+2} & \cdots & a_{2n} \\ & & \ddots & & & \ddots & \\ & & & & & & 1 \end{bmatrix}.$$

To show that

$$T_n(F) = H_1 \geq H_2 \geq \cdots \geq H_{n-1} \geq \{1\}$$

is a central series for $T_n(F)$, first note that if $I_n + A \in T_n(F)$ where

$$A = \begin{bmatrix} 0 & a_{12} & a_{13} & \cdots & a_{1n} \\ 0 & 0 & a_{23} & \cdots & a_{2n} \\ 0 & 0 & 0 & \cdots & a_{3n} \\ & & & \ddots & \\ & & & & 0 \end{bmatrix}$$

then $I_n + A^2 \in H_2, I_n + A^3 \in H_3, \ldots, I_n + A^n = I_n$. Also,

$$(I_n + A)^{-1} = I_n - A + A^2 - A^3 + \cdots + (-1)^{n-1} A^{n-1}.$$

Now let $I_n + B \in H_i$. Then we have

$$\begin{aligned} [I_n + A, I_n + B] &= (I_n + A)^{-1}(I_n + B)^{-1}(I_n + A)(I_n + B) \\ &= (I_n - A + \ldots)(I_n - B + \ldots)(I_n + A)(I_n + B) \\ &= I_n + AB + \text{ higher powers of } A \text{ and } B \\ &\in H_{i+1}. \end{aligned}$$

$T_n(\mathbb{Z}_p)$ consists of all upper triangular matrices with diagonal entries all 1 and arbitrary elements of $\mathbb{Z}_p$ in the $\frac{1}{2}n(n-1)$ positions above the main diagonal. It follows immediately that $|T_n(\mathbb{Z}_p)| = p^{\frac{1}{2}n(n-1)}$. But we know that

$$|\mathrm{SL}(n,p)| = \frac{1}{p-1} \prod_{i=0}^{n-1} (p^n - p^i)$$

(see question 1.18), and the highest power of $p$ that divides $|\mathrm{SL}(n,p)|$ is therefore $p^{\frac{1}{2}n(n-1)}$. Hence $T_n(\mathbb{Z}_p)$ is a Sylow $p$-subgroup of $\mathrm{SL}(n,p)$.

# Solutions to Chapter 3

**3.21**   $S_3$ can have non-trivial proper subgroups only of order 2 or 3. Only the Sylow 3–subgroup $\langle(123)\rangle$ is normal and so the only composition series is

$$S_3 > \langle(123)\rangle > \{1\}.$$

No Sylow subgroup of $S_4$ is normal, nor is any subgroup of order 2. Thus the only possible orders for proper non-trivial normal subgroups are 4, 6, and 12. But a subgroup of order 6 contains a Sylow 3–subgroup and so, if it is normal, contains all four Sylow 3–subgroups which is impossible since 4 does not divide 6. Any subgroup of order 12 contains all Sylow 3–subgroups and so is $A_4$ (observe that $A_n$ is generated by the 3-cycles). As $A_4$ has index 2, it is normal. It is easy to see that the only normal subgroup of order 4 is

$$V = \{(1), (12)(34), (13)(24), (14)(23)\}$$

and so the only composition series is

$$S_4 > A_4 > V > \{1\}.$$

$S_5$ has only one non-trivial proper normal subgroup and so

$$S_5 > A_5 > \{1\}$$

is the only composition series. To see this, we again use Sylow theory. If $\{1\} \neq N \lhd S_5$ and 5 divides $|N|$ then $N$ contains all six Sylow 5–subgroups of $S_5$ so $|N|$ is 30, 60, or 120. But each of these possibilities means that $N$ contains one and so all Sylow 3–subgroups of $S_5$ and so contains $A_5$ as above. Thus $N$ is either $A_5$ or $S_5$. Now if neither 3 nor 5 divides the order of $N$ then $N$ contains an element of order 2. But then, by conjugation, it contains at least 15 elements of order 2. Thus $A_5$ is the only possibile $N$. The above argument applies to subgroups of $A_5$ which is therefore simple, and $S_5$ has only one composition series.

If $n \geq 5$ then the only composition series of $S_n$ is

$$S_n > A_n > \{1\}.$$

**3.22**   As $A_{r+1}$ is a Sylow $q$–subgroup of $A_r$ and $A_{r+1} \lhd A_r$ we see that $A_r$ has only one Sylow $q$–subgroup. We show by induction on $i$ that $A_{r-i}$ has only one Sylow $q$–subgroup, namely $A_{r+1}$. Suppose that $A_{r-i+1}$ has only one Sylow $q$–subgroup $A_{r+1}$. Then $A_{r-i+1} \lhd A_{r-i}$ implies that if $g \in A_{r-i}$ then $g^{-1}A_{r+1}g \leq g^{-1}A_{r-i+1}g = A_{r-i+1}$ and so, by the assumption, $g^{-1}A_{r+1}g = A_{r+1}$. Thus $A_{r+1} \triangleleft A_{r-i}$. and so is the only Sylow $q$-subgroup of $A_{r-i}$. Hence, by induction, $A_{r+1} \triangleleft A_1 = G$. Thus $A_{r+1}$, and similarly $B_{s+1}$, are normal subgroups of $G$.

For $g \in A_{r+1}$ and $h \in B_{s+1}$ we have

$$g^{-1}h^{-1}gh \in A_{r+1} \cap B_{s+1} = \{1\}.$$

Thus $A_{r+1}$ and $B_{s+1}$ generate their direct product and the order of this shows that $A_{r+1}B_{s+1} = G$.

# Solutions to Chapter 4

**4.1**  The first part is achieved by a standard matrix reduction using the elementary operations of the forms

(a) add an *integer* multiple of one row/column to another;
(b) interchange two rows/columns;
(c) multiply a row/column by $-1$.

In this case the reduction begins and ends as follows :

$$\begin{bmatrix} 37 & 27 & 47 \\ 52 & 37 & 67 \\ 59 & 44 & 74 \end{bmatrix} \sim \cdots \sim \begin{bmatrix} 1 & 0 & 0 \\ 0 & 35 & 0 \\ 0 & 0 & 0 \end{bmatrix}.$$

This then shows that

$$G \simeq \langle\, x, y, z \mid x = 1,\ y^{35} = 1,\ z^0 = 1 \rangle \simeq C_{35} \times C_\infty.$$

If we now add the relation $a^3 b^2 c^4 = 1$ to those of $G$ then a corresponding matrix reduction gives

$$\begin{bmatrix} 37 & 27 & 47 \\ 52 & 37 & 67 \\ 59 & 44 & 74 \\ 3 & 2 & 4 \end{bmatrix} \sim \cdots \sim \begin{bmatrix} 1 & 0 & 0 \\ 0 & 7 & 0 \\ 0 & 0 & 0 \\ 0 & 0 & 0 \end{bmatrix}.$$

Thus adding the relation $a^3 b^2 c^4 = 1$ to the relations of $G$ changes $G$ to the group $C_7 \times C_\infty$. Hence the relation $a^3 b^2 c^4 = 1$ cannot hold in $G$. However, it follows immediately from the first two relations of $G$ that

$a^{15}b^{10}c^{20} = 1$ and so the order of $a^3b^2c^4$ divides 5. As the order is not 1, it must then be 5.

From the second and third relations it is clear that $(abc)^7 = 1$ and so $abc$ has order 1 or 7. Adding the relation $abc = 1$ to those of $G$ gives a matrix that reduces to

$$\begin{bmatrix} 1 & 0 & 0 \\ 0 & 5 & 0 \\ 0 & 0 & 0 \\ 0 & 0 & 0 \end{bmatrix},$$

which corresponds to the group $C_5 \times C_\infty$. Thus we see that $abc \neq 1$ in $G$ and that consequently $abc$ has order 7.

Now in an abelian group, if $x$ has order $m$ and $y$ has order $n$ then $xy$ has order l.c.m.$(m, n)$. Thus we deduce that

$$a^4b^3c^5 = a^3b^2c^4 \cdot abc$$

has order $5 \cdot 7 = 35$.

4.2      (a) The relation matrix for $G$ reduces as follows :

$$\begin{bmatrix} 2 & 3 & 6 \\ 4 & 9 & 4 \end{bmatrix} \sim \cdots \sim \begin{bmatrix} 1 & 0 & 0 \\ 0 & -2 & 0 \end{bmatrix}.$$

Consequently, $G \simeq \langle x, y, z \mid x = y^{-2} = z^0 = 1 \rangle \simeq C_2 \times C_\infty$.

     (b) The relation matrix for $G$ reduces as follows :

$$\begin{bmatrix} 2 & 3 & 6 \\ 4 & 9 & 4 \\ 3 & 3 & 2 \end{bmatrix} \sim \cdots \sim \begin{bmatrix} 1 & 0 & 0 \\ 0 & 1 & 0 \\ 0 & 0 & -66 \end{bmatrix}.$$

Consequently $G \simeq C_{66}$.

4.3      From $abab^2 = 1$ we have $bab = b^{-1}a^{-1}$. Hence $babab = 1$ and so $bab^{-1}a^{-1} = 1$, from which it follows that $G$ is abelian.

$G$ is the infinite cyclic group. In fact, a relation matrix for the abelian group $G$ is

$$\begin{bmatrix} 2 & 3 \end{bmatrix} \sim \begin{bmatrix} 2 & 1 \end{bmatrix} \sim \begin{bmatrix} 0 & 1 \end{bmatrix}.$$

Alternatively, it is easy to see that $G = \langle ab \rangle$ since $b = (ab)^{-2}$ and $a = (ab)^3$.

**4.4** Suppose that $g, h \in G$ have orders $m, n$ respectively. Then $g^{-1}$ has order $m$ and $gh$ has order that divides $mn$. Thus the non-empty subset $T$ is closed under multiplication and taking inverses, so it is a subgroup of $G$.

If $g \in Q$ and $h \in T \setminus \{1\}$ then $gh \in Q$. Thus, if $Q$ is a subgroup, we have $h = g^{-1}.gh \in Q$. It follows that a necessary condition for $Q$ to be a subgroup is that $T = \{1\}$. This condition implies that $Q = G$ and so is also sufficient.

Given a prime $p$, the element $f$ of $G$ defined by $f(p) = 1$ and $f(q) = 0$ for $q \in \Pi \setminus \{p\}$ is an element of order $p$.

The element $g$ of $G$ defined by $g(q) = 1$ for all $q \in \Pi$ has infinite order.

Let $f \in G$ be such that $f(p) \neq 0$ for only finitely many $p \in \Pi$. Suppose that $\{p_1, \ldots, p_n\}$ is the finite subset of $\Pi$ on which $f$ takes non-zero values. Then if $P = p_1 p_2 \cdots p_n$ it is readily seen that the order of $f$ divides $P$ (and in fact is equal to $P$), so $P$ has finite order.

Conversely, suppose that $f \in G$ has order $n$ say, so that $nf = 0$. We show that if $f(p) \neq 0$ for $p \in \Pi$ then $p$ must divide $n$. This will complete the proof since $n$ can have only finitely many distinct prime divisors. Now $f(p) \neq 0$ and $f(p) \in \mathbb{Z}_p$ imply that $f(p)$ has order $p$. But since $nf = 0$ we have $nf(p) = 0$, and so $p$ divides $n$.

**4.5** The relation matrix for $G$ is

$$\begin{bmatrix} n & m & m \\ m & n & m \\ m & m & n \end{bmatrix}.$$

This reduces as follows :

$$\begin{bmatrix} n & m & m \\ m & n & m \\ m & m & n \end{bmatrix} \sim \begin{bmatrix} n & m-n & m-n \\ m & n-m & 0 \\ m & 0 & n-m \end{bmatrix}$$

$$\sim \begin{bmatrix} 2m+n & 0 & 0 \\ m & n-m & 0 \\ m & 0 & n-m \end{bmatrix}.$$

The determinant is zero if and only if $m = n$ or $2m = -n$, whence the result follows.

If $G$ is perfect then $2m + n = 1$ and $n - m = 1$, so $m = 0$ and $n = 1$. Thus

$$G = \langle\, a, b, c \mid a = 1,\ b = 1,\ c = 1 \,\rangle$$

which is the trivial group.

*4.6* The relation matrix for $G$ is

$$\begin{bmatrix} 1 & 3 \\ 4+2k & 2n+9 \end{bmatrix}.$$

For $G$ to be perfect, we require the determinant of this matrix to be $\pm 1$. Now the determinant is

$$\Delta = 2n+9 - 3(4+2k) = 2n - 3 - 6k.$$

Since, by hypothesis, $n$ is coprime to 6, there are two cases to consider.

(a) $n = 6m+1$. In this case $\Delta = 12m+2-3-6k$ and we can choose $k = 2m$ to obtain $\Delta = -1$.

(b) $n = 6m - 1$. In this case $\Delta = 12m - 5 - 6k$ and we can choose $k = 2m - 1$ to obtain $\Delta = 1$.

*4.7* The relation matrix is

$$\begin{bmatrix} n & 0 \\ 3 & 1 \\ n+9 & 4 \end{bmatrix} \sim \begin{bmatrix} n & 0 \\ 3 & 1 \\ n & 1 \end{bmatrix} \sim \begin{bmatrix} n & 0 \\ 3 & 0 \\ 0 & 1 \end{bmatrix}.$$

Hence $G/G'$ is cyclic, of order h.c.f.$(3, n)$. Therefore $G/G' \simeq C_3$ if $n$ is divisible by 3, and is trivial if $n$ is coprime to 3.

*4.8* We have

$$V = S^{-1}R^{-y}$$
$$W = TV^{-z} = T(R^y S)^z$$
$$X = W^t U = (T(R^y S)^z)^t U$$

and hence

$$G = \langle\, R, S, T, U \mid R^x = S^a T^b U^c, \ (R^y S)^y = T^a U^d,$$
$$(T(R^y S)^z)^z = U^a, \ ((T(R^y S)^z)^t U)^t = 1 \,\rangle.$$

The relation matrix for $G/G'$ is therefore

$$M = \begin{bmatrix} x & -a & -b & -c \\ y^2 & y & -a & -d \\ yz^2 & z^2 & z & -a \\ yzt^2 & zt^2 & t^2 & t \end{bmatrix}.$$

We can simplify this relation matrix using elementary row or column operations over $\mathbf{Z}$. Add $-y$ times column 2 to column 1, then $-z$ times column 3 to column 2, then $-t$ times column 4 to column 3, and we obtain

$$M \sim \begin{bmatrix} x + ay & -a + bz & -b + ct & -c \\ 0 & y + az & -a + dt & -d \\ 0 & 0 & z + at & -a \\ 0 & 0 & 0 & t \end{bmatrix}.$$

Now $|G/G'| = \det M = (x + ay)(y + az)(z + at)t$. Hence $G/G'$ is finite if and only if $(x + ay)(y + az)(z + at)t \neq 0$.

(a) $|G/G'| = 1$ requires

$$x + ay = \pm 1, \; y + az = \pm 1, \; z + at = \pm 1, \; t = \pm 1$$

so we can take, for example, $a = 0, t = x = y = z = 1$.

(b) Take, for example, $t = 16, x = y = z = 1, a = b = c = d = 0$.

(c) Take, for example, $t = 2, z = 4, y = 8, x = 1, a = b = c = d = 0$.

**4.9** To see that $G = \langle a_1, a_2 \rangle$ we use induction. Suppose that $a_i \in \langle a_1, a_2 \rangle$ for all $i < n$. Then $a_n = a_{n-1} a_{n-2}$ shows that $a_n \in \langle a_1, a_2 \rangle$. Since $a_1, a_2 \in \langle a_1, a_2 \rangle$ it follows that $a_i \in \langle a_1, a_2 \rangle$ for $1 \le i \le 2m$ and so $G = \langle a_1, a_2 \rangle$.

To see that $f_{n-1} f_{n+1} - f_n^2 = (-1)^n$ we again use induction. The result is readily seen to hold for $n = 2$. Now

$$
\begin{aligned}
f_{n-1} f_{n+1} - f_n^2 &= f_{n-1}(2f_{n-1} + f_{n-2}) - (f_{n-2} + f_{n-1})^2 \\
&= 2f_{n-1}^2 + f_{n-1} f_{n-2} - f_{n-2}^2 - f_{n-1}^2 - 2f_{n-2} f_{n-1} \\
&= f_{n-1}^2 - f_{n-2}(f_{n-2} + f_{n-1}) \\
&= f_{n-1}^2 - f_{n-2} f_n \\
&= -(-1)^{n-1} \\
&= (-1)^n,
\end{aligned}
$$

the penultimate equality resulting from the inductive hypothesis.

Now in $G/G'$ the relation $a_{i+2} = a_{i+1} a_i$ allows us to write

$$a_i = a_1^{f_{i-2}} a_2^{f_{i-1}}.$$

Substituting these values of $a_i$ into $a_i = a_{i+2} a_{i+m+1}$ we obtain only two relations. From $i = 1$ we obtain

$$a_1^{f_m} a_2^{1+f_{m+1}} = 1,$$

and from $i = 2$ we obtain

$$a_1^{1+f_{m+1}} a_2^{1+f_{m+2}} = 1.$$

Hence we see that a relation matrix of $G/G'$ is

$$\begin{bmatrix} f_m & 1+f_{m+1} \\ 1+f_{m+1} & 1+f_{m+2} \end{bmatrix} \sim \begin{bmatrix} f_m & 1+f_{m+1} \\ 1+f_{m-1} & f_m \end{bmatrix}.$$

Consequently,

$$|G/G'| = |f_m^2 - (1+f_{m+1})(1+f_{m-1})|$$

from which the result follows on using the equalities

$$f_{m+1}f_{m-1} - f_m^2 = (-1)^m \quad \text{and} \quad g_m = f_{m+1} + f_{m-1}.$$

*4.10*   Since $b^2 = a^{2^{n-2}}$ is a relation of both $G$ and $H$ we need only show that the relation $(ab)^2 = b^2$ holds in $G$ to see that the relations of $G$ imply those of $H$. But clearly $bab^{-1} = a^{-1}$ implies $abab^{-1} = 1$, i.e. $(ab)^2 = b^2$.

To show that the relations of $H$ imply those of $G$, we need only show that $bab^{-1} = a^{-1}$ and $a^{2^{n-1}} = 1$ hold in $H$. Now $b^2 = (ab)^2$ implies $bab^{-1} = a^{-1}$ immediately. Raise $bab^{-1} = a^{-1}$ to the power $2^{n-2}$ to obtain

$$ba^{2^{n-2}}b^{-1} = a^{-2^{n-2}}.$$

But since $a^{2^{n-2}} = b^2$ we see that $a^{2^{n-2}}$ commutes with $b$. It follows that $a^{2^{n-2}} = a^{-2^{n-2}}$, i.e. $a^{2^{n-1}} = 1$.

*4.11*   It is easy to see that $H$ and $K$ are isomorphic. For, eliminating $c$ from the presentation of $K$ by setting $c = ab$ gives the presentation for $H$.

We now show that the relations of $H$ are consequences of those of $G$. In fact,

$$\begin{aligned} bab &= b^2 a^3 && \text{since } ab = ba^3 \\ &= a^2 a^3 && \text{since } b^2 = a^2 \\ &= a && \text{since } a^4 = 1. \end{aligned}$$

Also,

$$\begin{aligned} aba &= ba^3 a && \text{since } ab = ba^3 \\ &= b && \text{since } a^4 = 1. \end{aligned}$$

Next we show that the relations of $G$ can be deduced from those of $H$. In fact,

$$a^2 = abab \quad \text{since } a = bab$$
$$= b^2 \quad\quad \text{since } aba = b,$$

and

$$ab = bab^2 \quad \text{since } a = bab$$
$$= ba^3 \quad\quad \text{since } b^2 = a^2.$$

Finally, $b = aba = ba^3a = ba^4$ so $a^4 = 1$ as required.

Since the matrices

$$a = \begin{bmatrix} 0 & 1 \\ -1 & 0 \end{bmatrix}, \quad b = \begin{bmatrix} 0 & i \\ i & 0 \end{bmatrix}$$

generate a group of order 8, the last part of the question is routine.

4.12    Let $K = \langle a_1, \ldots, a_{n-1} \rangle$. Then setting $L = \langle a_n \rangle$ we have $G = KL$, and since $L \le Z(G)$ we must have $K$ normal in $G$. Now $G/K \simeq L$, an abelian group, so $G' \le K$. But since $a_n \in G'$ we have $L \le K$. Hence $KL = K$ and so $G = K$ as required.

Take as presentation for $Q_8$

$$\langle a, b \mid a^4 = 1, \ a^2 = b^2, \ ab = ba^3 \rangle.$$

Since $H/A \simeq Q_8$ and $A \le Z(H) \cap H'$ we have, by the above result, that $H = \langle \alpha, \beta \rangle$ and $\alpha^4 \in A, \alpha^2\beta^{-2} \in A, \alpha\beta^{-1}\alpha\beta \in A$. Now $[\alpha, \beta] = \alpha^{-2}a$ where $a \in A$ and so $[\alpha, \beta]$ commutes with $\alpha$ since $a \in Z(H)$. But $\alpha^2\beta^{-2} \in A$ so $[\alpha, \beta] = \beta^{-2}a'$ where $a' \in A$ and so $[\alpha, \beta]$ commutes with $\beta$. Thus $H' = \langle [\alpha, \beta] \rangle$. However, $\alpha^2$ commutes with $\beta$ since $\alpha^2\beta^{-2} \in A \le Z(H)$ and so we have

$$[\alpha, \beta]^2 = [\alpha, \beta]\alpha^{-1}\beta^{-1}\alpha\beta$$
$$= \alpha^{-1}[\alpha, \beta]\beta^{-1}\alpha\beta$$
$$= \alpha^{-2}\beta^{-1}\alpha^2\beta$$
$$= 1.$$

Therefore $H' \simeq C_2$. But $A \le H'$ and so, since $H/A \simeq Q_8$, we have $H/H' \simeq C_2 \times C_2$ whence $|H| = 8$ and $A = \{1\}$ as required.

**4.13**  We have $x^2 = y^2xy^{-1}$ and so
$$1 = x^8 = (y^2xy^{-1})^4 = (y.yx.y^{-1})^4 = y(yx)^4y^{-1}.$$
Hence $(yx)^4 = 1$ and so $(xy)^4 = 1$.
  Also, $y^2 = x^2yx^{-1}$ so
$$y^8 = (x.xy.x^{-1})^4 = x(xy)^4x^{-1} = xx^{-1} = 1.$$

**4.14**  Let $x = a^3b$ and $y = (a^2b)^{-1}$. Then since $xy = a$ it is clear that $x$ and $y$ generate $G$.
  Writing $a = xy, b = (xy)^{-3}x = y^{-1}x^{-1}y^{-1}x^{-1}y^{-1}$ we obtain
$$G = \langle\, x,y \mid (xy)^7 = y^3 = x^2 = (xy(y^{-1}x^{-1}y^{-1}x^{-1}y^{-1})^5)^2 = 1 \,\rangle.$$
Now the final relation can be written in the form
$$(xyy^{-1}x^{-1}y^{-1}x^{-1}y^{-1}(y^{-1}x^{-1}y^{-1}x^{-1}y^{-1})^4)^2 = 1.$$
Taking the inverse of this we obtain
$$((yxyxy)^4yxy)^2 = 1.$$
Now conjugate by $y$ to get
$$((y^2xyx)^4y^2x)^2 = 1.$$
Since $y^3 = 1$ we have $y^2 = y^{-1}$ and the required form follows.

**4.15**  Substituting
$$c = ab,$$
$$d = bc = bab,$$
$$e = cd = ab^2ab$$
into the relations of $G$ we obtain (1) and (2).
  Now
$$b = ab^2aba$$
$$= ab^2ab^2abab^2ab \quad \text{by (1)}$$
$$= ab^5ab \quad \text{by (2)}.$$
Thus we have $b^5 = a^{-2}$. Also,
$$a^2 = babab^2aba$$
$$= bab^2 \quad \text{by (2)}.$$
Hence $b^{-5} = a^2 = bab^2$ and so $a = b^{-8}$ as required.
  Replacing $a$ by $b^{-8}$ in (1) and (2) produces $b^{11} = 1$ and $b^{22} = 1$, so $G$ is the cyclic group $C_{11}$.

**4.16** From $ab = b^2a = b.ba = b.a^2b = ba.ab$ we obtain $ba = 1$. Hence $a = b^{-1}$, and substituting into $ab = b^2a$ we obtain $b = 1$. Now substitute $b = 1$ in $ba = a^2b$ to obtain $a = 1$. Hence $G$ is the trivial group.

Consider now $G_n$. The relation

$$a^i b^{n^i} a^{-i} = b^{(n+1)^i}$$

holds for $i = 1$. Also, assuming this equality we have that

$$
\begin{aligned}
a^{i+1} b^{n^{i+1}} a^{-(i+1)} &= a(a^i b^{n^i} a^{-i})^n a^{-1} \\
&= a b^{n(n+1)^i} a^{-1} \\
&= (ab^n a^{-1})^{(n+1)^i} \\
&= (b^{n+1})^{(n+1)^i} \\
&= b^{(n+1)^{i+1}},
\end{aligned}
$$

whence the result follows by induction.

Taking $i = n$ we obtain from the above

$$a^n b^{n^n} a^{-n} = b^{(n+1)^n}.$$

It follows that

$$ba^n b^{n^n} a^{-n} b^{-1} = b^{(n+1)^n}$$

and hence that

(1) $$a^{n+1} b^{n^n} a^{-(n+1)} = b^{(n+1)^n}.$$

But taking $i = n + 1$ gives

$$a^{n+1} b^{n^{n+1}} a^{-(n+1)} = b^{(n+1)^{n+1}}$$

and so, raising (1) to the power $n$ we obtain

$$b^{n(n+1)^n} = b^{(n+1)^{n+1}}$$

from which it follows that $b^{(n+1)^n} = 1$. Substituting this into (1) now produces $b^{n^n} = 1$. Since $n^n$ is coprime to $(n+1)^n$ we then have that $b = 1$, so $a = 1$ also and $G_n$ is trivial.

**4.17** Let $H = \langle a, b \rangle$. Then $H$ is not abelian. To see this, let

$$A = \begin{bmatrix} 1 & 1 \\ 5 & 6 \end{bmatrix}, \qquad B = \begin{bmatrix} 6 & 2 \\ 6 & 1 \end{bmatrix}$$

so that we have

$$ab = \natural(A), \qquad ba = \natural(B).$$

Now $\natural(A) \neq \natural(B)$ since otherwise we would have $AB^{-1} = \pm I_2$ which is false.

A simple computation shows that

$$a^2 = b^4 = (ab)^2 = \natural \begin{bmatrix} 6 & 0 \\ 0 & 6 \end{bmatrix}$$

where $\natural \begin{bmatrix} 6 & 0 \\ 0 & 6 \end{bmatrix}$ is the identity of PSL$(2,7)$. Hence $H$ is an image of

$$D_8 = \langle a, b \mid a^2 = b^4 = (ab)^2 = 1 \rangle$$

using von Dyck's theorem. However, any proper image of $D_8$ has order 1, 2, or 4, and so is abelian. This then shows that $H \simeq D_8$.

**4.18** By question 1.18, SL$(2,3)$ has order 24. The elements of GL$(2,3)$ have determinant 1 or 2, and the same argument that counts the elements of determinant 1 clearly shows that there are 24 elements with determinant 2. Hence

$$|\text{GL}(2,3)| = 48.$$

Since SL$(2,3)$ has index 2 in GL$(2,3)$ it must be normal and contain the derived group. But

$$\begin{bmatrix} -1 & 0 \\ 0 & 1 \end{bmatrix}\begin{bmatrix} 1 & 1 \\ 0 & 1 \end{bmatrix}\begin{bmatrix} -1 & 0 \\ 0 & 1 \end{bmatrix}\begin{bmatrix} 1 & -1 \\ 0 & 1 \end{bmatrix} = \begin{bmatrix} 1 & 1 \\ 0 & 1 \end{bmatrix}$$

and

$$\begin{bmatrix} -1 & 0 \\ 0 & 1 \end{bmatrix}\begin{bmatrix} 1 & 0 \\ 1 & 1 \end{bmatrix}\begin{bmatrix} -1 & 0 \\ 0 & 1 \end{bmatrix}\begin{bmatrix} 1 & 0 \\ -1 & 1 \end{bmatrix} = \begin{bmatrix} 1 & 0 \\ 1 & 1 \end{bmatrix}.$$

It is now straightforward to check that

$$\text{SL}(2,3) = \left\langle \begin{bmatrix} 1 & 1 \\ 0 & 1 \end{bmatrix}, \begin{bmatrix} 1 & 0 \\ 1 & 1 \end{bmatrix} \right\rangle$$

and so $\mathrm{SL}(2,3)$ is the derived group of $\mathrm{GL}(2,3)$.

We now have

$$H = \langle\, a,b,c,\vartheta \mid ab = c,\ bc = a,\ ca = b,\ \vartheta^3 = 1,\ \vartheta^{-1}a\vartheta = b,$$
$$\vartheta^{-1}b\vartheta = c,\ \vartheta^{-1}c\vartheta = a \,\rangle.$$

A little trial and error soon produces the correspondence

$$a \longleftrightarrow \begin{bmatrix} 0 & -1 \\ 1 & 0 \end{bmatrix}, \qquad b \longleftrightarrow \begin{bmatrix} 1 & 1 \\ 1 & -1 \end{bmatrix},$$

$$c \longleftrightarrow \begin{bmatrix} -1 & 1 \\ 1 & 1 \end{bmatrix}, \qquad \vartheta \longleftrightarrow \begin{bmatrix} 1 & -1 \\ 0 & 1 \end{bmatrix}.$$

This then shows that $H \simeq \mathrm{SL}(2,3)$ since each has order 24.

The presentation for $H$ may be simplified by eliminating $b$ and $c$ using the fifth and sixth relations, to obtain

$$H = \langle\, a,\vartheta \mid \vartheta a\vartheta a^{-1}\vartheta a^{-1} = 1,\ \vartheta^3 = 1 \,\rangle.$$

Now $H/H' \simeq C_3$ and is generated by $\vartheta$. Hence $H' \simeq Q_8$ and the derived group of $Q_8$ is $\langle a^2 \rangle \simeq C_2$ with quotient group $C_2 \times C_2$.

The derived series of $\mathrm{GL}(2,3)$ is now seen to be

$$
\begin{array}{cl}
 & \bullet\ \mathrm{GL}(2,3) \\
C_2 & \Big| \\
 & \bullet\ \mathrm{SL}(2,3) = H \\
C_3 & \Big| \\
 & \bullet\ H' = Q_8 \\
C_2 \times C_2 & \Big| \\
 & \bullet\ C_2 \\
 & \Big| \\
 & \bullet\ \{1\}
\end{array}
$$

**4.19**  Let $T = \langle\, ab, ab^{-1}ab \,\rangle \le H$. Then

$$ab^{-1} = ab^{-1}ab(ab)^{-1} \in T$$

and so $b = b^{-2} = (ab)^{-1}ab^{-1} \in T$ and $a = ab.b^{-1} \in T$. Hence $T = H$ and so $H$ is generated by $ab$ and $ab^{-1}ab$.

Let $M = \langle (ab)^n \rangle \le H$. We show first that $M$ is central in $H$. Since $H = \langle ab, ab^{-1}ab \rangle$ it is sufficient for this purpose to show that $[(ab)^n, ab] = 1$ and $[(ab)^n, ab^{-1}ab] = 1$. Now the first of these is clear, and the second follows on substituting $(ab^{-1}ab)^k$ for $(ab)^n$. Since also

$$(ab)^n = (ab^{-1}ab)^k = (a^{-1}b^{-1}ab)^k \in H'$$

we see that $M \subseteq H'$ as required.

*4.20*   To show tht the commutator $[a, b]$ is in the centre of $G$, it suffices to prove that it commutes with each of the generators $a$ and $b$. Now

$$
\begin{aligned}
a^{-1}[a, b]a &= a^{-1}(a^{-1}b^{-1}ab)a \\
&= a(b^{-1}a)ba \quad \text{since } a^3 = 1 \text{ gives } a^{-2} = a \\
&= a(a^{-1}ba^{-1}b)ba \quad \text{since } (b^{-1}a)^3 = 1 \text{ gives } b^{-1}a = (a^{-1}b)^2 \\
&= ba^{-1}b^{-1}a \quad \text{since } b^2 = b^{-1} \\
&= b(baba)a \quad \text{since } a^{-1}b^{-1} = (ba)^2 \\
&= b^{-1}aba^{-1} \quad \text{since } b^2 = b^{-1}, a^2 = a^{-1} \\
&= a^{-1}ba^{-1}bba^{-1} \quad \text{since } b^{-1}a = (a^{-1}b)^2 \\
&= a^{-1}ba^{-1}b^{-1}a^{-1} \quad \text{since } b^2 = b^{-1} \\
&= a^{-1}bbabaa^{-1} \quad \text{since } a^{-1}b^{-1} = (ba)^2 \\
&= a^{-1}b^{-1}ab \quad \text{since } b^2 = b^{-1} \\
&= [a, b],
\end{aligned}
$$

and so we have that $[a, b]$ commutes with $a$. A similar proof shows that $[a, b]$ commutes with $b$.

Since the commutator $[a, b]$ is thus in $Z(G)$ we have that $G/Z(G)$ is abelian and $Z(G)$ is abelian. But an abelian group must be finite whenever it is generated by a finite number of elements and is such that, for all $x$ and a fixed $n$, $x^n = 1$. Hence $Z(G)$ is finite, $G/Z(G)$ is finite, and so $G$ is finite.

*4.21*   We have $y^a x y^b = x y^{-c} x$ and $y^{-c} x y^{-b} = x y^a x$, so

(1) $$y^a(xy^{b-c}x)y^{-b} = xy^{-c}x.xy^a x = xy^{a-c}x.$$

Similarly,

(2) $$y^b(xy^{c-a}x)y^{-c} = xy^{b-a}x$$

and

(3) $$y^c(xy^{a-b}x)y^{-a} = xy^{c-b}x.$$

From (1) and (3) we have

$$y^{2a}(xy^{b-a}x)y^{-b-c} = xy^{a-c}x$$

and using (2) we obtain

$$y^{2a+b}(xy^{c-a}x)y^{-b-2c} = xy^{a-c}x.$$

Hence

$$y^{2a+b}(y^{b+2c}xy^{a-c}xy^{-2a-b})y^{-b-2c} = xy^{a-c}x$$

and so

$$[y^{2(a+b+c)}, xy^{a-c}x] = 1.$$

Similarly

$$[y^{2(a+b+c)}, xy^{c-b}x] = 1.$$

If h.c.f.$(a-c, c-b) = 1$ we have $\lambda(a-c) + \mu(c-b) = 1$ and then

$$[y^{2(a+b+c)}, (xy^{a-c}x)^\lambda(xy^{c-b}x)^\mu] = 1$$

and so $[y^{2(a+b+c)}, xyx] = 1$. But now

$$(xyx)^a = xy^ax = y^{-c}xy^{-b}$$

so $[y^{2(a+b+c)}, y^{-c}xy^{-b}] = 1$, giving $[y^{2(a+b+c)}, x] = 1$ as required.

4.22  Call $xt^{m+1} = t^2x^2$ relation (1) and $xt^2xtx^2t = 1$ relation (2). From (1) we obtain

$$xt^{m+1}x^{-1} = t^2x$$

so, squaring and using (2),

$$xt^{2m+2}x^{-1} = t^2xt^2x = tx^{-2}t^{-1}.$$

This then establishes (a).
   From (2) we have $t^{-2} = (xtx)^2$ so $[t^2, xtx] = 1$, which is (b).
   By (1) and (b) we have

$$t^m = t^{-1}x^{-1}t^2x^2 = (xt^2x^{-1}t^{-1}x^{-1})x^2.$$

91

Hence
$$xt^2x^{-1} = t^m x^{-1} t$$
$$= t^{2m+1} x^{-2} t^{-1} \quad \text{by (1)}$$
$$= t^{2m} t x^{-2} t^{-1}$$
$$= t^{2m}(xt^{2m+2}x^{-1}) \quad \text{by (a)}.$$

This establishes both (c) and (d).
   Finally,
$$t^{2m} = t^{2m}(xt^{m+1}x^{-2}t^{-2})$$
$$= (xt^{m+1})t^{-2m}(x^{-2}t^{-2})$$
$$= (xt^{m+1}x^{-2}t^{-2})t^{-2m},$$

so $t^{4m} = 1$ and hence, using (d), $[t^{2m}, x] = 1$ which shows that $t^{2m} \in Z(G)$.

4.23   Since $\begin{bmatrix} a & b \\ c & 0 \end{bmatrix} \in \mathrm{SL}(2, \mathbb{Z})$ we have $bc = -1$ and so $b = \pm 1$. Suppose that $b = -1$. Then we have

$$\begin{bmatrix} a & b \\ c & 0 \end{bmatrix} = \begin{bmatrix} a & -1 \\ 1 & 0 \end{bmatrix} = \begin{bmatrix} 1 & a \\ 0 & 1 \end{bmatrix}\begin{bmatrix} 0 & -1 \\ 1 & 0 \end{bmatrix} = s^a t \in \langle s, t \rangle.$$

Suppose now that $b = 1$. Then if

$$m = \begin{bmatrix} a & 1 \\ -1 & 0 \end{bmatrix}$$

we have, by the above,

$$t^2 m = \begin{bmatrix} -1 & 0 \\ 0 & -1 \end{bmatrix}\begin{bmatrix} a & 1 \\ -1 & 0 \end{bmatrix} = \begin{bmatrix} -a & -1 \\ 1 & 0 \end{bmatrix} \in \langle s, t \rangle.$$

Consequently, $m \in \langle s, t \rangle$.
   Now choose $n$ such that $|b + nd| < |d|$. Then if

$$m = \begin{bmatrix} a & b \\ c & d \end{bmatrix}$$

we have

$$s^n m = \begin{bmatrix} 1 & n \\ 0 & 1 \end{bmatrix}\begin{bmatrix} a & b \\ c & d \end{bmatrix} = \begin{bmatrix} a + nc & b + nd \\ c & d \end{bmatrix}$$

we have

$$s^n m = \begin{bmatrix} 1 & n \\ 0 & 1 \end{bmatrix} \begin{bmatrix} a & b \\ c & d \end{bmatrix} = \begin{bmatrix} a+nc & b+nd \\ c & d \end{bmatrix}$$

and hence

$$ts^n m = \begin{bmatrix} 0 & -1 \\ 1 & 0 \end{bmatrix} \begin{bmatrix} a+nc & b+nd \\ c & d \end{bmatrix} = \begin{bmatrix} -c & -d \\ a+nc & b+nd \end{bmatrix}.$$

Now $ts^n m \in \langle s,t \rangle$ by induction, so $m \in \langle s,t \rangle$.

Writing

$$u = st = \begin{bmatrix} 1 & -1 \\ 1 & 0 \end{bmatrix},$$

we have

$$u^3 = \begin{bmatrix} -1 & 0 \\ 0 & -1 \end{bmatrix}, \qquad t^2 = \begin{bmatrix} -1 & 0 \\ 0 & -1 \end{bmatrix}$$

and so $\bar{u}^3 = \bar{t}^2 = \bar{I}$ in $\mathrm{PSL}(2,\mathbb{Z})$.

To show that we can assume that $w$ has the given form, note that conjugation will transform words with different beginnings or endings to this form.

That $ut = -s$ follows by a simple matrix multiplication. Thus, putting $v = u^{-1}t$, we have

$$w = u^{\pm 1} t u^{\pm 1} \cdots u^{\pm 1} t = \pm \cdots s^{n_i} v^{n_{i+1}} s^{n_{i+2}} v^{n_{i+3}} \cdots$$

where $\ldots, n_i, n_{i+1}, \ldots$ are positive integers. But

$$s^\alpha = \begin{bmatrix} 1 & \alpha \\ 0 & 1 \end{bmatrix}, \qquad v^\beta = \begin{bmatrix} 1 & 0 \\ \beta & 1 \end{bmatrix}$$

so any product of the above form is a matrix whose entries are all non-negative and, provided both $s$ and $t$ occur, the trace exceeds 2. Hence $w \neq \pm I$ and so

$$\mathrm{PSL}(2,\mathbb{Z}) = \langle \bar{u}, \bar{t} \mid \bar{u}^3 = \bar{t}^2 = \bar{I} \rangle,$$

since we have shown that no further non-trivial relations can hold.

# Test paper 1

Time allowed : 3 hours
(Allocate 20 marks for each question)

1     Prove that the centre of a group of order $p^n$ is non-trivial. Let $K$ be a finite group and let $H$ be a subgroup of $K$. If $P_1$ is a Sylow $p$–subgroup of $H$, explain why $P_1 \leq P_2$ for some Sylow $p$–subgroup $P_2$ of $K$.

     Now suppose that $H$ satisfies the condition that if $h \in H$ and $h \neq 1$ then $N_K(h) \leq H$. By considering the centre of $P_2$, or otherwise, show that $P_1 = P_2$.

     Deduce that h.c.f.$(|H|, |K : H|) = 1$.

2     Let the quaternion group $Q_8$ be given by the presentation

$$Q_8 = \langle\, a, b \mid a^4 = 1,\ a^2 = b^2,\ aba = b \,\rangle.$$

Show that the mappings $\alpha, \beta$ defined by

$$\alpha(a) = ab, \quad \alpha(b) = a$$
$$\beta(a) = b, \quad \beta(b) = a$$

extend to automorphisms of $Q_8$.

     Let $G = \langle\, \alpha, \beta \,\rangle$. Prove that $G$ is a group of order 24 isomorphic to $S_4$. Show also that

$$G \simeq \operatorname{Aut} Q_8.$$

3     Suppose that $A$ is a set of generators of a group $G$ and that $H$ is a proper subgroup of $G$. Given an element $a$ of $A$ not belonging to $H$, let $B$ be the set obtained from $A$ by replacing each $x \in A \cap H$ by $ax$. Show

that $B$ is a set of generators of $G$. If $A$ is finite and has $n$ elements, show that $B$ has at most $n$ elements.

Deduce that

(i) if $G$ has $n$ generators then it has $n$ generators lying outside a given proper subgroup;

(ii) if $H$ is a proper subgroup of $G$ then $G \setminus H$ generates $G$.

**4**    (a) Prove that every subgroup $H$ of an (additive) cyclic group $G$ is cyclic and show that if $a$ is a generator of $G$ and $H$ has index $n$ then $na$ is a generator of $H$. If the order of $G$ is $m$, show that $b$ is also a generator of $G$ if and only if $b = ra$ and $a = sb$ for some integers $r, s$ both coprime to $m$. Deduce that if, in addition, $G$ is a $p$–group and $d$ is any generator of $H$ then there is a generator $c$ of $G$ such that $nc = d$.

(b) Let $p$ be a fixed prime. Suppose that $G$ is an additive abelian group with the property that it contains precisely one subgroup $H_\alpha$ of order $p^\alpha$ for each $\alpha$, and no other subgroups. Show that $H_\alpha \subseteq H_{\alpha+1}$ and that $H_\alpha$ is cyclic. Deduce using (a) that there are generators $x_0 = 0, x_1, \ldots, x_\alpha, \ldots$ of $H_0 = \{0\}, H_1, \ldots, H_\alpha, \ldots$ such that $px_{\alpha+1} = x_\alpha$ for every $\alpha$.

Consider the additive group

$$Q = \{ \frac{\beta}{p^\alpha} \mid \beta, \alpha \in \mathbb{Z},\ \alpha \geq 0 \}$$

of rational numbers. Show that $\beta/p^\alpha \mapsto \beta x_\alpha$ describes a group morphism from $Q$ to $G$ and deduce that $G \simeq Q/\mathbb{Z}$.

**5**    Express the abelian group

$$G = \langle\, x, y, z \mid x^6 y^6 z^9 = x^2 y^4 z^3 = x^4 y^4 z^6,\ xy = yx,\ yz = zy,\ zx = xz \,\rangle$$

as a direct product of cyclic groups.

Suppose that $\alpha$ is a morphism from $G$ such that $\operatorname{Im} \alpha$ is of odd order. Show that $\operatorname{Im} \alpha$ is cyclic.

Let $H$ be a group in which $g^2 = 1$ for every $g \in H$. Show that $H$ is abelian. If the order of $H$ is finite show that it is $2^n$ for some positive integer $n$.

# Test paper 2

Time allowed : 3 hours
(Allocate 20 marks for each question)

**1**     Let $G$ be a finite group of order $p^m n$ where $p$ is a prime that is coprime to $n$. What do the three Sylow theorems tell you about the $p$–subgroups of $G$?

Show, by using induction on the order or otherwise, that a maximal subgroup of a finite $p$–group $P$ is normal in $P$.

Supose that $G$ has at least three Sylow $p$–subgroups $P_1, P_2, P_3$ where $P_1 \cap P_2$ and $P_2 \cap P_3$ are maximal subgroups of index $p$ in $P_2$. Show that $P_1 = (hk)^{-1} P_3 hk$ where $h \in \mathcal{N}_G(P_2 \cap P_3)$ and $k \in \mathcal{N}_G(P_1 \cap P_2)$.

**2**     Prove that every subgroup of a nilpotent group is subnormal. Deduce that a maximal subgroup of a nilpotent group is normal.

Let $G$ be a group in which every finitely generated subgroup is nilpotent, and let $M$ be a maximal subgroup of $G$. Suppose that $M$ is not normal in $G$. Prove that there is an element $x$ of $G'$ with $x \notin M$. Writing $x = \prod_{i=1}^{n}[y_i, z_i]$, prove that $\{x, y_i, z_i \mid i = 1, \ldots, n\}$ is contained in a subgroup $H$ of $G$ where

$$H = \langle\, x, a_1, \ldots, a_m \mid a_i \in M, i = 1, \ldots, m \,\rangle.$$

Let $A = \langle\, a_1, \ldots, a_m \,\rangle$ and let $L$ be maximal in $H$ with respect to the property that $A \leq L$ and $x \notin L$. Show that $L$ is a maximal subgroup of $H$, that $x \in H'$, and that $H' \leq L$.

Deduce from the above that a maximal subgroup of a group in which every finitely generated subgroup is nilpotent is normal.

**3**  Find permutations $x, y \in A_5$ with $x^2 = 1, y^3 = 1, (xy)^5 = 1$. Show that $A_5$ has a presentation

$$\langle \, x, y \mid x^2 = y^3 = (xy)^5 = 1 \, \rangle.$$

By considering the matrices

$$X = \begin{bmatrix} 10 & 8 \\ 8 & 1 \end{bmatrix}, \quad Y = \begin{bmatrix} 5 & 7 \\ 5 & 5 \end{bmatrix}$$

in $SL(2, 11)$, find a subgroup of $PSL(2, 11)$ that is isomorphic to $A_5$.

**4**  An additive (resp. multiplicative) abelian group $G$ is said to be *divisible* if for every $x \in G$ and every non-zero integer $n$ there exists $y \in G$ with $ny = x$ (resp. $y^n = x$).

Prove that the additive group of rationals is divisible, and that so also is the multiplicative group of complex numbers of modulus 1.

Show that no proper subgroup of the rationals is divisible.

**5**  Show that the group $K$ with presentation

$$\langle \, a, b, c, d \mid ab = d, \; bc = a, \; cd = b, \; da = c \, \rangle$$

is cyclic of order 5. Hence or otherwise find the order of the group with presentation

$$L = \langle \, a, b, c, d \mid ab = d, \; ad = c, \; bc = a, \; cd = b, \; da = c \, \rangle.$$

Show that the group $M$ with presentation

$$\langle \, a, b, c \mid abcabc = a, \; bcabca = b, \; cabcab = c \, \rangle$$

is cyclic and determine its order.

# Test paper 3

Time allowed : 3 hours
(Allocate 20 marks per question)

**1**  Let $G$ be a finite group and let $p$ be a prime dividing the order of $G$. Let $P_1, \ldots, P_r$ be the Sylow $p$–subgroups of $G$. Show that the mapping $\vartheta_g$ from $\{P_1, \ldots, P_r\}$ to itself defined by

$$\vartheta_g(P_i) = gP_ig^{-1}$$

is a bijection. Show also that the mapping $\Theta$ from $G$ to the group of bijections on $\{P_1, \ldots, P_r\}$ given by $\Theta(g) = \vartheta_g$ is a morphism whose kernel is the largest normal subgroup of $G$ that is contained in the normaliser in $G$ of a Sylow $p$–subgroup.

Let $G$ be a group of order 168 which has no non-trivial proper normal subgroups. Show that $G$ cannot be represented non-trivially as a permutation group on fewer than seven letters. Show that $G$ can be represented as a permutation group on eight letters.

**2**  Let $G$ be a nilpotent group and $H$ a normal abelian subgroup of $G$ with the property that $H$ is not properly contained in any normal abelian subgroup of $G$. Prove that $H = \{g \in G \mid (\forall h \in H)\, [g, h] = 1\}$.

Deduce that $H$ is not properly contained in any abelian subgroup of $G$ and that $\operatorname{Aut} G$ contains a subgroup isomorphic to $G/H$.

**3**  Show that if $p$ is prime then $\natural : \mathbf{Z} \to \mathbf{Z}/p\mathbf{Z}$ induces a morphism from $G^* = \mathrm{SL}(2, \mathbf{Z})$ to the group $G_p^* = \mathrm{SL}(2, \mathbf{Z}_p)$.

Given that in both $G^*$ and $G_p^*$ the centre is the subgroup generated by

$$\begin{bmatrix} -1 & 0 \\ 0 & -1 \end{bmatrix},$$

explain why the above morphism induces a morphism

$$\vartheta_p : G^*/Z(G^*) \to G_p^*/Z(G_p^*).$$

Show that conjugation by the element

$$\begin{bmatrix} 0 & 1 \\ -1 & 0 \end{bmatrix}$$

induces an automorphism $\tau$ of order 2 of $\mathrm{Ker}\, \vartheta_p$. Prove also that

$$\begin{bmatrix} 0 & 1 \\ -1 & 0 \end{bmatrix} \notin \mathrm{Ker}\, \vartheta_p$$

and that $\tau(x) = x$ implies $x = 1$. By considering the matrices

$$\begin{bmatrix} 1 & p \\ 0 & 1 \end{bmatrix} \quad \text{and} \quad \begin{bmatrix} 1 & 0 \\ p & 1 \end{bmatrix}$$

show that $\mathrm{Ker}\, \vartheta_p$ is not abelian.

Let $G$ be a finite group and let $\tau$ be an automorphism of $G$ such that $\tau^2 = 1$ and $\tau(x) = x$ implies $x = 1$. Show that if $x^{-1}\tau(x) = y^{-1}\tau(y)$ then $x = y$. Deduce that $\tau$ inverts every element of $G$. Hence prove that $G$ is an abelian group of odd order.

4    Prove that every quotient group of a nilpotent group is nilpotent, and that every finite $p$-group is nilpotent.

Find the order of the group

$$G_n = \langle\, a, b \mid a^{2^n} = b^2 = (ab)^2 = 1 \,\rangle.$$

Prove that $G_n/Z(G_n) \simeq G_{n-1}$. Hence show that $G_n$ is nilpotent of class $n$.

5    Express the abelian group

$$\langle\, a, b, c \mid a^{14}b^3c^{11} = a^8b^{-3}c^{11} = a^3b^3 = 1,\ ab = ba,\ bc = cb,\ ca = ac \,\rangle$$

as a direct product of cyclic groups. Show that the subgroup of elements of finite order is cyclic and find a generator for it.

# Test paper 4

Time allowed : 3 hours
(Allocate 20 marks for each question)

1    If $G$ is a finite group and $H, K$ are subgroups of $G$ prove that

$$|HK| = \frac{|H||K|}{|H \cap K|}.$$

If $G$ is a group of order 48 with more than one Sylow 2–subgroup, find the possible number of Sylow 2–subgroups. If $P_1, P_2$ are distinct Sylow 2–subgroups, prove that $|P_1 \cap P_2| = 8$. Show also that $P_1 P_2 \subseteq N_G(P_1 \cap P_2)$. By considering $|N_G(P_1 \cap P_2)|$ show that $P_1 \cap P_2$ is a normal subgroup of $G$.

Hence show that any group of order 48 has a proper non-trivial normal subgroup.

2    Show that the number of elements in a conjugacy class in a finite $p$–group is a power of $p$. Deduce that a non-trivial finite $p$–group has a non-trivial centre.

Show that if $P$ is a non-trivial finite $p$–group then $P$ contains subgroups $P_1, \ldots, P_k$ such that

$$P = P_1 \supset P_2 \supset \cdots \supset P_k = \{1\},$$

each $P_i$ is a normal subgroup of $P$, and $|P_i : P_{i+1}| = p$ for $i = 1, \ldots, k-1$.

3    Let $G$ be a group with $G' \leq Z(G)$. Prove that, for all $x, y \in G$ and all integers $n \geq 1$,

$$x^n y^n = (xy)^n [x, y]^{\frac{1}{2}n(n-1)}.$$

Suppose now that $G = \langle x, y \rangle$. Prove that if $g \in G$ then $g = x^a y^b [x, y]^c$ for some integers $a, b, c$.

Deduce that if $H$ is the subgroup of $\mathrm{SL}(3, \mathbb{Z})$ given by

$$\left\{ \begin{bmatrix} 1 & b & c \\ 0 & 1 & a \\ 0 & 0 & 1 \end{bmatrix} \;\middle|\; a, b, c \in \mathbb{Z} \right\}$$

then $\vartheta : H \to G$ described by

$$\vartheta \begin{bmatrix} 1 & b & c \\ 0 & 1 & a \\ 0 & 0 & 1 \end{bmatrix} = x^a y^b [x, y]^c$$

is a surjective group morphism. Deduce that $G$ is a quotient group of $H$.

4    If $H$ and $K$ are nilpotent groups prove that so also is $H \times K$. What is the class of $H \times K$ in terms of the classes of $H$ and $K$?

Let $M, N$ be normal subgroups of a group $G$. Prove that the mapping $G \to G/N \times G/M$ given by $g \mapsto (gN, gM)$ is a morphism. Hence show that if $G/N$ and $G/M$ are nilpotent then so also is $G/(N \cap M)$. What can you say about the class of $G/(N \cap M)$ in terms of the classes of $G/N$ and $G/M$?

5    Express the abelian group

$$G = \langle\, a, b, c \mid a^2 b^2 c^6 = a^4 b^6 c^4 = a^6 b^4 c^4 = 1, ab = ba, ac = ca, bc = cb \,\rangle$$

as a direct product of cyclic groups.

Find the number of elements of order 11 in $G$. Show that every element of order 11 in $G$ is of the form $a^{2\alpha} b^{2\beta} c^{2\gamma}$ for some integers $\alpha, \beta, \gamma$. Find an element of order 11 in $G$ and express it in terms of the generators $a, b, c$.